Praise for *On Gallows Down*

From treetop protests at the Newbury Bypass to the grand Highclere Estate, *On Gallows Down* is that rare thing: nature writing as political as it is personal.

Melissa Harrison, author of
The Stubborn Light of Things: A Nature Diary

A powerful personal and political journey through place that charts the profound influence we have on nature, and that nature has on us.

Rob Cowen, author of
Common Ground and *The Heeding*

An evocative and inspiring memoir which touches on environmental protest, family, motherhood and, most importantly, nature. Her passion for the natural world and especially birds, shines through in this wonderful book.

Claire Fuller, author of *Unsettled Ground*

Nicola Chester deserves many readers. *On Gallows Down* is an impassioned study of a contested landscape, which interrogates our attitudes towards land stewardship, ownership and living in the right relationship with both human and other-than-human neighbours. Charged with love and fire, *On Gallows Down* is a beautiful exploration of a much-mapped, multi-faceted landscape.

Katharine Norbury,
author of *The Fish Ladder*

Chester's writing has a lovely elasticity, dancing between wonder, introspection and anger as she moves from the particular to the universal…She belongs to the disappearing English, rural working class, and is intent on handing this baton to her three children, who play a part in the book. Chester also explores the familiar tension between wanting to write and being needed at home. The heady ecstasy of time carved out alone, in nature. The scrabble to earn a precarious living, and the insecurities of occupying a tied cottage. The idea of 'home' lies at the heart of this fierce, beautifully written, immersive book about one's place within the landscape.

Tessa Boase, author of
Etta Lemon: The Woman Who Saved the Birds

Nicola's passionate and enduring love of nature shines through every single word, paragraph and page of this book, as she seamlessly weaves memoir with stories of the landscape in which she is so deeply rooted that it seems to speak through her. Powerful, enlightening, dazzling, hopeful, *On Gallows Down* is a rare and precious gem – to be savoured, not rushed, and returned to again and again. My words cannot do this book justice – it simply needs to be read.

Brigit Strawbridge Howard, author of the
Wainwright-shortlisted *Dancing with Bees*

ON GALLOWS DOWN

Also by Nicola Chester

RSPB Spotlight: Otters (Bloomsbury, 2014)

ON GALLOWS DOWN

Place, Protest and Belonging

Nicola Chester

Chelsea Green Publishing
White River Junction, Vermont
London, UK

Copyright © 2021 by Nicola Chester.
All rights reserved.

No part of this book may be transmitted or reproduced in any form by any means without permission in writing from the publisher.

Developmental Editor: Muna Reyal
Project Manager: Alexander Bullett
Copy Editor: Caroline West
Proofreader: Laura Jorstad
Designer: Melissa Jacobson

Printed in the United Kingdom.
First printing September 2021.
10 9 8 7 6 5 4 3 2 1   21 22 23 24 25

ISBN: 978-1-64502-116-2

Chelsea Green Publishing
85 North Main Street, Suite 120
White River Junction, Vermont USA

Somerset House
London, UK

www.chelseagreen.com

For the flock, feather, nest and homecoming that is my family.

*To Martin, who saved me
from the man on the big white horse,
and to our children
– Billy, Evie and Rosie –
and the hill that raised them.*

*There are some heights in Wessex, shaped as if by a kindly hand
For thinking, dreaming, dying on, and at crises when I stand,
Say, on Ingpen Beacon eastward, or on Wylls-Neck westwardly,
I seem where I was before my birth, and after death may be.*

Thomas Hardy, 'Wessex Heights'
from *Satires of Circumstance: Lyrics and Reveries*

The place we occupy seems all the world

John Clare, 'The Shepherds Almost
Wonder Where They Dwell'

Contents

1	Bird in a Landscape	1
2	Common Land	8
3	Ten Thousand Trees	28
4	An Extinction of Nightingales	45
5	The Thing with Feathers: Fieldfaring	61
6	A Library of Landscape	73
7	Tenant, Tied	91
8	Home of Homes, A High, Clear Place	109
9	The Form of a Hare	124
10	Flint, Feather and Bone	139
11	On Gallows Down	153
12	The Green Plover's Nest	171
13	Rural Work: The Dew Pond on the Height	186
14	Watering	203
15	An Ecology of Love and Ruin	217
16	Kite Flying	229
	Acknowledgements	*240*

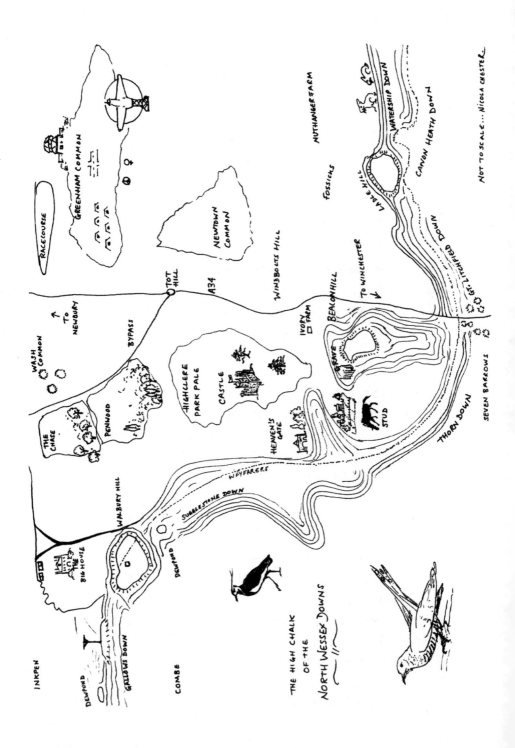

CHAPTER ONE

Bird in a Landscape

It is St George's Day, late April, two days shy of my birthday. The sky is the colour of a pheasant's egg and skylarks are singing against it at such a height I can't see them. A just-discernible shimmer of heat blurs the near horizon of orange gravel that marks the old runway of this former US airbase. I am sitting on my hands on top of an old American fire hydrant, its once-smooth sides speckled with rust and yellow and red paint curled and crusted like lichen. I can't quite reach the ground and sit swinging my legs, a toe occasionally reaching a knobbly chunk of flint to kick away. I think I've been stood up.

The view to the southwest is all curves. The open-mouthed caves of the old cruise-missile bunkers replicate the smooth, lyrical, whaleback arc of the chalk hills beyond and their ancient procession of round and long barrows and hillforts. The striking green contours of bare, richly flowered and springy turf take in White Hill, Watership Down, Ladle Hill and Great Litchfield Down. On the other side of the hidden A34 are the 'elephant graves' of Seven Barrows on the Highclere Estate where British aviation pioneer Geoffrey de Havilland made his first flight in a homemade aeroplane in 1909. Above Seven Barrows is the graceful dome of Beacon

Hill, crowned with its Iron Age hillfort. At its centre lie the railings and grave of the (reputedly mummified) 5th Earl of Carnarvon, co-discoverer of the tomb of the Egyptian boy pharaoh, Tutankhamun. Only from that hill can I see the old, isolated, flint-and-brick estate cottage at Highclere that was once my home. I still have the key to its blue-liveried front door on a piece of matching ribbon by the side of my bed. I like to weigh its cool heaviness in my hand and for my fingers to remember the satisfying grate it made when turned in a lock the size of a child's shoebox. From there, the hills roll on to where I live now, within an afternoon's walk along the ridgeway.

If I lower my gaze again to the foreground, the old bunkers, built to hold nuclear warheads as well as withstand a strike from one, are now long-haired, softened relics that have become a part of the narrative-landscape of this place. Steep-sided, flat-roofed, grassed all over, they seem part Neolithic long barrow, part natural chalk downland; half-built pyramids for a pharaoh or shallower versions of the prehistoric flat-topped cone of Silbury Hill not many miles from here. Their doorless entrances are wide mouths, open and raised obediently to the sky, all six singing the same long note: say 'ahhh'. They are empty vessels, waiting for a spoonful of something.

I am sitting in the middle of what was formerly RAF Greenham Common, a military airbase in West Berkshire, Southern England, a few miles south of the town of Newbury. Now just Greenham Common, it is being restored to what it had been for thousands of years – a thousand acres of open heath on a gravel table, with wet, wooded, primeval gullies that run off it like creases in a tablecloth. Greenham Common, almost as I know it, was created millennia ago, when the little, benign chalk streams of the Kennet and the Enborne – which gave my schoolhouses two of their names – were mighty rivers that washed out the flints from chalky subsoil overlying sunset-coloured clays and sands.

Greenham Common is a place of big skies and wide, cloud-reflecting pools. The joyful sense of freedom and open space I felt 21 years ago when the fences came down is renewed every time I come back. Back then, that feeling had a lot to do with the Common's military and social history, its place in local culture and how I remembered it before the fences as a child. Yet now, that sense has become permanent, emanating from the place itself like an excitable shiver of heat haze on the low horizon. A force of nature. This high, wild, gravelly plateau, romantically bleak one day and a riot of crackling warmth and colour on another, is our own wild moor. Richard Mabey, the father of modern writing on nature, wrote about Greenham Common in 1993 when it was decommissioned and up for sale: 'It will be heartwarming if the place can become a common ground for humans as well as wildlife...a powerful symbol...complete with grazing animals, ponds and a few rusting relics to remind future generations of what this place once was.' The sense of happy incredulity that this actually came to pass does not diminish.

Before then, the birds and plants were mostly nameless familiars. I knew their touch, sight, sound and smell – the memories they engendered, the feelings they inspired – but not always their names. They were there. They were abstract. But they were part of my personal history. We used to exercise the riding-school ponies bareback along the paths and tracks in the school holidays, over purple heather or in snow, until one day an eight-foot chain-link fence cut right across the path. Another path, presumably an attempt at compensation, was diverted and, bizarrely, concreted. The hoofprints we made that day in the fresh, almost-set cement registered a protest and are still there. But now the path ends nowhere again; itself having become a relic, like the old control tower, the fire hydrants and the 'fire plane', a rusty facsimile of an aeroplane that lies supported above a round, shallow pond which would have been filled with burning fuel for training purposes. The odd concrete post or angled section of rusted razor-wire fence seems innocuous, even naïve now. They wouldn't keep a rabbit out. Yet I remember the very real threats of a 'shoot to kill' policy if the inner enclosure was breached.

Apart from the birdsong and the occasional popping of gorse and the patter of explosively ejected seeds falling, the Common is quiet now. I check my watch and decide to give it another ten minutes, shifting myself on my fire hydrant. The wood shakes itself of cherry blossom.

Greenham Common is an ancient place, never otherwise enclosed. It had remained under traditional common use, by local cottagers and smallholders, for grazing, fuel and other gathered materials, right up until it was requisitioned in 1941 for a wartime airfield. But instead of being returned to common land, as expected, after the war ended, the UK government overruled considerable local opposition and protest and it became a United States airbase for heavy bombers. By 1980, the government agreed for it to be a base for American nuclear-armed cruise missiles.

The Common expirates a largely undocumented historical aura of commoners and the rural poor as well as Romanies, travellers, thieves, picnickers, refugees and armies, but it has a more recent retro feel, too. For me, no doubt because it's a childhood haunt full of touchstones and memories; but also because of its missile hangars with their 1980s mullet hairdos and other Cold War accoutrements, the 1950s-style American fire hydrants and 1940s ammunition stores hidden deep in a bluebell wood, with its abandoned, crumbling, concrete stalls. Silver-washed fritillaries the colour of strong tea float down to rest on the old brickwork and, sometimes, you can happen upon a lizard basking, curvy like the letter 'S', or a diamond-studded adder, and mistake them for an ornate hinge, a rusted bracket, the imprint of a mountain-bike tyre in grey earth or an old piece of heavy-duty rope. Sometimes little ringed plovers gather round the pools and there are woodlark, green-winged orchids and nightingales beyond the fire plane. And on a day when the clouds build and loose like billowy galleons over the downs and the ponies and commoners' brindled, dappled cattle graze among the stands of gorse (or furze, depending on which side of the border you're from), it's a part of North Wessex that Thomas Hardy would recognise, as long as you keep the view to the southwest and the smells of the pickle factory to your east. It has a lived-in, handed-down, given-back feeling. Pre-worn, nostalgic but freshly aired. Like a faded piece of vintage cloth, washed and made into a new dress, it is quietly jubilant. It's a landscape that connects, threads through and links my own narrative, intrinsically. In the year 2000, having been derequisitioned and reinstated as common land once more, Greenham Common was officially reopened to the public. I watched a short-eared owl cruise down the old runway where once there were Vulcan Bombers.

In ten minutes, I must leave to pick the children up from school. The one I'm waiting for has still not arrived. I am no lovelorn teenager in the all-consuming, hopeful, fearful flush of first love, but there is a growing, gnawing emptiness I can barely contemplate: that everything hinges on this moment and that I will break if it does not happen. The one I wait for always arrives within the two days either side of my birthday. I have been holding a candle for him all my known life, lighting it every spring, and each year the anxiety that he will not make it increases. The knowledge of what he – and generations of him – have to go through each time he comes and goes increases year on year, making our meeting seem all the more unlikely. That thought almost prostrates me with grief. I am waiting for a bird. I am waiting for the cuckoo.

It hasn't always been the cuckoo I've waited so hard for. Once, for a while, there was another heartbreaker bird. I come here each year just to hear that too, now; over beyond the fire plane, because now that is the only place I can find it. And this is important. Because I need to know that the cuckoo I want so much to hear is – as near as possible – 'my' cuckoo; the cuckoo (or another generation of it) that belongs to this landscape as much as I do, that is still coming back, still returning. I need to know that the cuckoo is still coming home.

Between the fire hydrant and the fire plane lie the remnants of the old airstrip, the concrete cross of its middle all that remains of the longest runway in Europe. The rest of it has been broken up into rubble and is the reason I feel that having the Common back is only some sort of consolation. The rubble now lies, recycled, in a narrow, nine-mile stretch (the same length as the old, unravelled perimeter fence) three miles from here, dividing and curving between my old homes, under the Newbury Bypass.

Time was up for me on Greenham. It was another ten days before I finally heard the cuckoo. Carrying out washing to the line, I heard it call from the wood across the field. That two-syllable note that echoes down the centuries with the quality of a woodwind instrument. '*Cuc-koo, Cuc-koo*'. It stopped me in my tracks and before I could help it a sharp intake of breath became a sob. I felt as though my legs had been knocked from under me and I had fallen, winded, only to realise that I'm okay, that it's okay. There is the hot sting of tears and I wonder, when did I get like this? About a bird? When did it ever begin to mean so much?

CHAPTER TWO

Common Land

When did it start, this love affair with place and nature? This need, this delight and revelment, this affirmation, wonder and comfort? This longing? Long before I did, certainly. Flawed, sophisticated, ridiculous – capable of great love, beauty, cruelty and destruction – we are hardwired to nature. It turns out, it is in our interest to be. It is a love that provides consolation, even as it is lost; and therefore, along our road to planetary ruin, we can find a way back to some kind of human evolution that includes the earth we stand on, the air we breathe and all the living, interconnected things that we manage to simultaneously hold in awe and ignore. An ecology of love and hope.

But it also provides us with a map and waymarkers for home, a way into a place and a place to start from. It gives us ground, and a grid reference for our place of greater safety.

I am fortunate and grateful that I had parents and grandparents who knew the importance of outdoors: of walks, weather and picnics, time alone, freedom and dirt. But who also knew the importance of connection and meaning. In small childhood, I remember names of places: the homes we lived in as a family in Hampshire and West Berkshire. Denmead, in the Meon Valley near Petersfield where I was born;

Winchester, where my brother Ian was born three years later; Tilehurst and Pangbourne on the rural outskirts of Reading in Berkshire; my paternal grandparents' home in the village of Rothwell, Northamptonshire; and Portsdown Hill, the big, wild chalk hill that rose above my maternal grandparents' council-estate home in Paulsgrove, Portsmouth – and of these places I remember chalk streams, ironstone amber and woods. Front doors, perhaps, family of course, dogs and dog walks, the smell and sting of nettles, an inexplicable fear of long grass and a terror of spiders. I remember trees, butterflies, birds, soldier beetles, water voles – and horses, wherever I could find them. Particularly with the landscape I explored around my grandparents' homes, there was a sense of something 'other' than quite belonging; a certain anxiety and discomfort in being an 'outsider' to children in those places, and it set me off exploring the countryside whenever I could, alone. And this was exciting. Here, I belonged. Here, I found confidence.

There was so much joy and a personal, inner freedom outdoors, in nature. It was the first thing I looked for, wherever I was – an alleyway between houses, bordered with 'weeds'; a plum or crab apple tree cascading down on the 'unowned' side of the fence; an unfenced piece of rough ground, the ford of a stream, birds riding along with me through a bus window – these things were not owned by anyone, or even, I found, noticed by everyone. So, I claimed them, inwardly, secretly, joyfully, as my own.

―――――

We moved to Pangbourne, a Thames-side village not far from Reading, when I was eight, from one fire-brigade-owned house (that came with Dad's job as a firefighter) to another on a housing estate. The estate abutted onto allotments, fields and the little River Pang. It was my 'Anahorish', my 'place of clear water' where I first acknowledged what I can only describe as a yearning and attachment for nature and home. The two were absolutely entwined. There, I was wild and free and happy. Like most children of the time, my parents turned me out of doors and I came back when I was hungry. Sometimes alone, with friends, or in a gang, I played in the farmland and water meadows off the footpaths, known as 'The Moors'. I tamed cows, fished for tiddlers and sheltered from the worst rain showers in a cowpat-emulsioned Second World War pillbox. When the meadows flooded, between the clear, chalk Sulham Brook and River Pang, we'd wade in wellies.

When Dad moved to a different fire station to work in Newbury, he and Mum were able to buy a small terraced house at Greenham, between the woods and Newbury Racecourse. When we moved from Pangbourne to Greenham, it felt like a coming away of skin.

I realised instinctively, then, how much the nature I knew offered me. I had felt a part of it and it was most certainly a part of me. I could be alone in nature, quite happily (and needed to be often, in fact, as a shy and introverted child). It sustained me endlessly, through physical and imaginative exploration, but also through enquiry into the world and through emotional reflection and attachment. It was absorbing. Exciting, even thrilling at times. Being in it made me a better, more engaged, calmer person. I already knew it wasn't something anyone could *own* but I felt, with absolute conviction, that I belonged to it, like the birds belonged to the sky – but to the earth and the trees too.

It was also apparent to me, even then, that such a strong, early attachment and identity with all things wild would be tested. My access to it – particularly to specific places – was not within my control. I could be moved away or prevented from being there, and beloved things could be destroyed or torn down without addressing my need for them. I was, somehow, on the wrong side of a fence I had trouble acknowledging was there at all.

From that point on, those themes – the importance of access to nature, the loss and protection of it, historical, imaginative and literary connection, belonging and a sense of 'ownership' and freedom through nature – became a path through the wilderness of everything else. What else could I do, but walk it?

In moving house, I felt I had been plucked from one benign, lush, sheltering 'moor' and dropped in another, very different one, not much more than a dozen miles away: though it was not referred to by name as a 'moor', it was, in every imaginative sense, exactly that – but wilder, bigger, scratchier.

———

It is hard to think that certain spots, certain landscapes, do not absorb human history, strong passions and lives in ways less tangible than the physical, and exhale it. Greenham Common exudes this through every pore. Every skylark or woodlark rising will be singing a song listened and attended to by people connected to moments in time in this place. A woodlark's *'allelu-lu-luia'* through the centuries becomes a hymn for this common ground, an individual weight of meaning we might all recognise and claim. A song of home.

The heathland on Greenham Common predates the Bronze Age barrows echoed by the old nuclear silos, as well as the agriculture that surrounds it. Heathland is an almost-accidental man-made environment, millennia old, which provided people with subsistence living and sustenance; more often than not when there was none to be had elsewhere. On infertile, poor, sandy and acidic soils of little or no value, it was common land for those that had none, or too little. A place to graze animals, to harvest fuel and fodder, and to extract sand and gravel – it was an extension of home, a wilder kind of allotment and, sometimes, a home for the homeless.

Prior to 1830, William Cobbett (among other things, farmer, journalist and champion of the rural poor) took several of his celebrated and documented *Rural Rides* in the area. But he deemed Greenham's lonely vista 'a villainous tract of rascally heath'. 'Where the place? Upon the Heath. There to meet with Macbeth,' chime the three Shakespearean witches. Indeed. But in the early 1900s, Victor Bonham-Carter (author, farmer, publisher) spent some of his childhood living at Greenham, my new home, and knew it more intimately, describing it as 'a mighty wilderness… threaded by a single dust road.'

Greenham Common has a long and sporadic military history, first recorded in the English Civil War. In 1643, the first Battle of Newbury (which took place just three miles away) was one of the decisive conflicts culminating in a victory for the Parliamentary forces. Troops trained, camped, rallied and marched from here and continued to do so at various intervals in time: some 6,000 redcoats were billeted on the heath during the Jacobite Rebellion in Scotland and probably marched from Greenham to fight in the Battle of Culloden in 1746. From 1859 right up until the First World War, up to 20,000 troops regularly camped there, often on their way to the nearby military training ground of Salisbury Plain. Local people would visit to cheer the troops, watch manoeuvres and listen to the regimental bands. These armies left little trace, until the Second World War. Indeed, in 1915, Charles Rothschild, a keen entomologist, naturalist and wealthy banker, drew up a well-researched list of 284 of the country's best and most important sites for wildlife to be 'preserved in perpetuity'. Greenham Common was among them, until it was declared 'ruined' by the Second World War.

Then, much of the common land of this relatively high, flat, gravel table was requisitioned, enclosed, developed and shared between the Royal Air Force and United States Army Air Forces. On 6 June 1944, General Eisenhower famously addressed the Allied Expeditionary Forces from Greenham Common as they took part in the D-Day landings, announcing, 'The eyes of the world are upon you. The hopes and prayers of liberty-loving people everywhere march with you.' Just weeks earlier, intense and earnest training by the British 9th Battalion, the Parachute Regiment, had taken place a few miles along the downs from Greenham, under the hill above the house where I now live. Paratroopers trained for their lives to disable and capture the Merville Battery gun emplacement ahead of those troops landing on the Normandy beaches on 6 June 1944. As the troops built up on Greenham, Royal Engineers spent seven days constructing a full-scale replica of the gun emplacement, its anti-tank ditches, barbed wire and minefields among the fields and woods below this steep chalk escarpment. Nine Para carried out nine practice assaults, four of them at night, using live ammunition. The remote downland village of Inkpen, where I now live, was shut off from its hill, and its footpaths closed down for the duration. Security was taken so seriously that members of the Women's Auxiliary Air Force were co-opted to 'test' the reliability of the paratroopers to keep their mission secret, attempting to elicit information via a presumed 'honey trap'. Those men that failed the test were denied their last 48 hours of freedom before flying to France. The cornfields are still known as 'Trenchfields' locally. Now and then, Will Carter, our current near-neighbour and farmer on the Estate where we live, turns up armaments in his tractor, and miles from any watercourse, a reinforced concrete pillbox (a type of defensive guard post that could be fired from within, which is more usually seen beside canals and waterways) nestles under the hill, itself bunker-like, with its roguish toupée of ryegrass and bramble.

After the war, it was hoped that Greenham Common would once again be returned to the people, but by April 1951, the perceived threat from the Soviet Union during the Cold War rose enough for the Ministry of Defence to hand sole occupation of the Common over to the United States Air Force to be used as a Strategic Air Command base.

Despite (or perhaps because of) growing tension, several big international military airshows were held on the Common between 1973 and 1983. Now boasting the longest runway in Europe, the build-up would begin two weeks before, with planes arriving from all over the world, flying heart-thumpingly low over both the primary and secondary school. It wasn't a time for schoolwork. The aeroplanes' bolted underbellies blotted out the sun, their great engines rattling the single-pane windows of our Portakabin classrooms in their soft-puttied, prefabricated frames. We waited always for 'the big one', listening and measuring the approaching roar of engines and the trembling of desks against the wrath of our teachers – the American cargo plane we knew (in hushed, whispered awe) as the Galaxy Starlifter was always worth the dash to the window and, often, right out the door. (It's a habit I've not lost – on hearing the low bass thump of a Chinook from Salisbury Plain come over the house.) With the exception perhaps of the Vulcan Bomber, little compared to the exhilaration of a Galaxy Starlifter coming in to land right over your school. But these were also the supersized cargo planes that brought the cruise missiles in, putting an end forever to the airshows as well as our freedom of the Common.

Before the bombs, the Peace Women came. In August 1981, four women from Cardiff, deeply opposed to the UK's agreement to house American nuclear weapons, organised an (almost) all-women march from their hometown in Wales to Greenham Common. They stopped off in Newbury the night before they reached the Common to buy padlocks and chains to secure themselves to the perimeter fence. The women represented all ages, and several were young mothers with their children in pushchairs. Many stayed to form the permanent Women's Peace Camp, staged at different gates around the Common that remained, with women – and always just women – coming and going for the next 19 years. Local people had been used to spirited and gutsy protests from the separate, but linked Campaign for Nuclear Disarmament; had encountered marches against nearby Aldermaston's Atomic Weapons Establishment; and witnessed increasingly angry, inspired protests. In December 1982, 30,000 women 'embraced the base', linking arms around its razor-wired perimeter in an extraordinary hug, initiated by a chain letter, of all things. In April 1983, 70,000 protestors linked arms to make an astonishing 14-mile human chain, linking Greenham Common to Aldermaston and the nuclear ordnance factory on Burghfield Common.

Britain had accepted 160 of America's nuclear weapons of mass destruction and, in November 1983, 96 of those were flown in to be housed on the airbase (100 if you counted the 'four spares'). These fully operational atomic cruise missiles were kept in the six specially constructed bunkers. Some 10 metres high, the bunkers were insulated with a layer of concrete, 2 metres thick, and a layer of sand, 3 metres deep, over a titanium plate, covered with tons of clay. They were designed to withstand a nuclear air-burst from above or a direct hit from a 2,500lb (1,100kg) conventional bomb. We watched them being built from the remaining tracks around the Common from a guarded distance, with the knowledge that each of the 96 missiles had the equivalent destructive power of 16 Hiroshima bombs.

It was a strange place to be, growing up in strange times. My hometown of Newbury appeared in the daily press as 'Nukebury'. My generation grew up knowing that the end of the world could come at the push of a button, at any moment, in the form of an imagined mushroom cloud and the terrible fallout that would follow.

Fear of the 'four-minute warning' – the promised sirens that would wail out across the country in the event of imminent nuclear attack – haunted the fringes of our consciousness. But where could we go to in four minutes? There were no nuclear fallout shelters. In music videos and on television, there were images of a white-out blast that would do for us all as we sat in our cars in the ultimate traffic jam. If my parents had any fear, they never showed it. I understood it was a ridiculous situation to be in and it was best not to think about it. Any questions were met by Dad's dismissive and practical gallows humour. As a very young man during the 1962 Cuban Missile Crisis, he'd been on the only Royal Navy ship in the Caribbean, facing oncoming Russian ships at the moment all-out nuclear war between Russia and the US seemed imminent. It was averted in the last few minutes of the eleventh hour. He left the navy when I was born. A career in the fire brigade followed.

Dad would reassure my brother and me that we were so close to Greenham and its missiles that if anything happened, we'd know nothing about it. It was a strange kind of comfort, but, into my teens, I accepted it, although I felt the cold, ashy taste of fear in my mouth when the sirens were tested. This must have happened at regular, published times because the reassuring 'just testing' always came from someone (usually a teacher during the school day), though it was never absolutely convincing. There was usually an uneasy pause; a very brief moment where everyone stopped and looked up towards the door or windows and thought about how far they could get in four minutes, as the eerie wail wound itself up into the rise and fall of the siren.

In 1947 Chicago, the Bulletin of the Atomic Scientists devised and published the 'Doomsday Clock', a metaphorical clock face representing how close humanity is to global disaster (midnight) through our own doing. When Dad was in the Caribbean in 1962, the clock hands were put at seven minutes to midnight. In 1983, with cruise missiles in place in England, the clock hands were set at three minutes to midnight – not enough time for even a four-minute warning. Eight years later, the United States and Soviet Union signed the Strategic Arms Reduction Treaty, and the clock was turned back to 17 minutes to midnight. Then, the global disaster was nuclear war, but in 2015, eight years since the threat of climate change first influenced the hands of the clock, concerns around the lack of worldwide political action to address the climate emergency that was unfolding elevated the threat level on a par with the continued modernisation of nuclear weapons. The clock was set once again at 11.57.

But in 2020, the hands inched closer still. The two greatest threats to life on the planet, before factoring in the effects of a global pandemic, remain nuclear war or accident and man-made climate change. Scientists have now set the Doomsday Clock to just 100 seconds to midnight. The closest it has ever been. Like a horological sandwich board, the Doomsday Clock predicts that, unless we sort ourselves out, the end of the world is nigh.

Meanwhile, the further enclosure and sealing off of our Common, and the arrival of the American soldiers and their families, MOD Police and the Peace Women, were something I encountered daily. This was my home ground. A place I played and mooched about in, hung out with a friend and walked the dog. It was a place I spent time in on my own and where I (absent-mindedly, reflectively) studied the wildlife. I honed my more 'animal' skills and learnt to walk quietly as a roe deer; to listen, look, sniff, poke about and to leave little trace of having been there. I could walk over leaves so quietly; I could creep up on my friend. (Even now, at work in the school library, I can place my feet just so and walk unheard, apologetic as a muntjac, past readers in the reference section. In heels.)

Into my teens, weekends and school holidays were spent working at the riding stables, grooming, mucking out and exercising the horses, and taking customers out for hacks. The only place to ride was on the Common. In fact, the only way onto what remained of the Common and its wooded edges was to actually ride through the middle of one of the Peace Camps and along the narrow tracks beside the razor wire, past armed guards in towers. It was the first time I'd ever seen a gun. It seemed to me that everyone inside was holding one, pointing at the outside. The Peace Women were mostly treated viciously by locals. They were seen as a dirty, smelly inconvenience. Shops refused to serve them, pubs banned them, local householders refused requests for water. The camps, often because of constant evictions and the confiscation of property, were sometimes little more than hastily put-up 'bender tents' – sheets of plastic thrown over arches of hazel rods – and were considered an eyesore. Although most local people didn't want the US base there, they didn't want the women there either. There seemed to be a general consensus that they were unruly, untrustworthy, disrespectful troublemakers who should have been home with their families – and this upset me. It felt awkward to ride, high on a horse, through their constantly disrupted and devastated camp of broken tents, protest songs and cooking pots, and I felt embarrassed and ashamed by some of my peers' reactions which seemed to define the protestors by sexuality: they were all either 'lesbian outcasts', or sleeping with American soldiers, or had been abandoned by their husbands for being dirty and depraved. I was careful to be apologetic and cheery; it seemed to me we wanted the same thing – for the Americans, their missiles and the fences to go away and for the Common to be given back. I had little political awareness, but I resented the American soldiers for their perceived 'warmongering' and the way the Common had been taken, and at the same time, opened up to the world.

But the Americans held some fascination. Most were charming, bored and curious, and there was sport to be had. Here were we, half a dozen local teenage girls on the wild, free side of the Common, galloping about, often riding bareback with the wind in our hair and all the time in the world. We didn't have much in common really, other than the Common and our beloved horses; some borrowed, some owned. We never spoke at school. A couple of crucial years of secondary school might separate us, or our social background: among those of us from the council or racecourse estate were a vet's daughter, a doctor's daughter and a solicitor's daughter. We united around the horses, taunted and teased the soldiers, fearless and confident in our riding abilities, challenging their jeeps and motorbikes to hair-raising races across unforgiving gravel – and forgiving heather – all behind the safety of the fence.

Sometimes, late on summer evenings, a strange but familiar bird would strike up and, for a time, it took on a more ghostly, sickening mantle in my imagination. A bird, I was well aware, with a rich, wildly fanciful cultural history. Half-apparition already, it easily lent itself to the story my best friend Lesley and I made up, to frighten American soldiers and their teenage sons. We presumed American girls weren't allowed out. We never saw any. Our careful, curious relationship with the boys was complicated and conducted at arm's length. We were awed, half-envious, a little afraid and suspicious. They were confident, exotic, loud and brash. They told us tall, boastful tales and, in return, we told them ghost stories. In our own untutored ways, we were each trying to tell the other who we were and where we were coming from; to claim the ground we felt we owned (and that they had been thrust upon) as ours. We played to our strengths, each drawing on the only resources we had. We were the clever, rightful, witchy natives on the blasted heath, with all that history to hand. 'Have you heard old Goatsucker?' we would start. 'He's an old goat shepherd, long dead, who used to graze the common you're now on…he'll call you, "*Qui*ck, *qui*ck!" and then he plays this pipe…' In reality, old Goatsucker was a bird, a nightjar, a summer visitor that haunted the heath with its after-dusk, ventriloquial churring and '*coo*ick' calls. A bird that has lulled me to sleep on occasion with its sound of my mother's sewing machine. A bird that dances and claps its wings together just above the ground like a jerky marionette. Inquisitive and confiding, you can call it to you, conjure it up out of the night with a double-clap of your hands, so it flies upright from the heather with its odd puppetry, circling your head or hovering in front of your face like a dark, seraphic angel. The churring song comes disembodied and is the sound of a strange machine, generator-like, that stops and starts and fades, only to strike up again loudly when you are least expecting it.

There's nothing like a bit of beguiling bewitchery from local country girls to further unsettle the sleep-starved, homesick, overactive imaginations of soldiers on night patrol. Particularly when they are staring for hours at the coal-black darkness of the wild heath beyond the razor wire, outward from the sensory deprivation of their buzzing, hyper-lit military space. And particularly, too, when their immediate enemy on the other side was not Russia. Not the one they'd trained for, not one they could fight, but peaceful women. An enemy that had wile, humour, eloquence and the certainty they had moral right on their side, and who could disarm a soldier with a smile. They did not really know who we were, either the Peace Women or us local girls; nor did they know how to separate us, if that were even possible. There were times when we felt very much part of the same collective. But when the goats from nearby Bunker Farm escaped their own fences, as they often did, and the birds would start calling into the night, distant and near and distant again, it corroborated our 'story' and frightened them witless.

In a way, I was outside it all, an observer, as I was with everything else on the Common. But I was local. I was from here-ish, which nobody else seemed to be. What started as indignant, righteous, adolescent games became a way to feel in control and even a little respected or feared. But as I became a young adult, I frightened myself with my own stories. I sickened of the whole situation: the verbal abuse from all parties, the astonishing squalor the Peace Women endured, the military hardcore, the missiles, the fences, the nasty sexual insults and lewd 'come-ons' from men much older than my friend and me. Even when we were just passing by, not part of the protest at all, we were women outside the fence and therefore, it seemed, permitted targets. At this time too, in a cluster that mirrored the flight path and spread downwind from the runway, directly over the primary school and through the estate where we lived, there was an unexplained, almost four-fold increase in the number of leukaemia and cancer cases diagnosed – particularly, and tragically, in those aged between 0 and 24. My brother's young schoolfriend was one of the victims.

When I heard the nightjar for the first time again one summer, it was against an evening backdrop of a part of the Common you wouldn't want to be alone on; a lonely lane off the A339, where there was always torn pornography and condoms strewn about the layby. We'd stumbled across the astonishing sight of the military convoys training on what looked like the troughs and deep mud of a Salisbury Plain tank track, only the mud and water was orange from the natural iron in the gravel, and not milky from the chalk. The flints were flung, all the same. This was the Ground Launched Cruise Missile System. Its purpose was either to transport the missiles to Salisbury Plain for a pre-emptive attack or to launch from the concrete outside the bunkers within two minutes in defensive retaliation. I was never sure which was the worse human action. This was 'Mutually Assured Destruction', and the initials said it all. The convoys looked like something out of a futuristic American film, inspired by some dystopian comic book. Each convoy carried 16 unarmed missiles in four Transporter Erector Launchers (a name, surely, no woman would ever have come up with). Unmarked, in dark gunmetal and green, this monster tank nightmare was escorted by two launch control centres and 69 troops. They were meant to melt into the countryside. Each secret convoy, which plenty of people saw but all authorities denied, seemed dreamt up, imagined. On the local roads at night, when they rushed past the diminutive Mini Metros and Ford Fiestas, like some kind of apocalyptic ghost train, we were left wondering if we really saw it. Rumour and conspiracy theories were rife and there was a locally held belief that the convoy travelled from Greenham Common to nearby RAF Welford, another enormous munitions store just a handful of rural miles west of Newbury with access to the M4 motorway. Many believed the missiles were actually stored there and Greenham was just an enormous, visible bluff.

On this particular dark, sultry night of summer cloud cover, we watched the great freight road train handle the off-road training track when, above and below its noise, and seemingly all around and even inside my head, came the disembodied sound of the nightjar. But here, now, it suddenly felt all wrong. It jarred weirdly. My friends did not know what it was. They couldn't believe the sound came from a bird that hadn't stopped to breathe, to take a breath. I remember trying to demonstrate by whistling a long note, in and out, and it all ending in uneasy laughter. I began to doubt myself. The machines bounced and splashed their way round the circular track like some demonic, unpeopled fairground ride. And in this place of strewn pornography, where a friend and I had once been flashed at by a group of motorbikers, I frightened myself with my own ghost story. My much-loved, pastoral shepherd's bird had become a bird that made the sound of a Geiger counter. A 'song' that rose and fell with the sickening breathing in and out of the four-minute warning sirens.

In a moment, all my post-adolescent, young-adult fears – sex and death, trust, self-belief and sanity – converged on those strange, mechanical, churring notes that emanated from a puppet bird and contaminated earth. A sound that jangled and pulsed through my nervous system, stopping and then starting from somewhere else deep inside the woods – and my own body. In a feeling that has recurred at intervals through my life, I felt disembodied, adrift from the world. As if I, or everyone else, were not really there. I think I cried. I remember going home to play Kate Bush's 1980 single, 'Breathing', quietly on my bedroom record player; her lyrics, from the perspective of an unborn child breathing in plutonium fallout during a nuclear blast, were matched to a symphony and rhythm of in-and-out breaths, all haunted by the sound of the nightjar, whose own notes still rose and fell in my head, in time with my own breathing. I listened, and tried to come to terms with it all.

All the time, the Peace Women remained an edifying force. Each camp at each gate named for the colours of the rainbow, when reality meant that their camp, torn down and destroyed by bailiffs on a daily basis, was often little more than a brown mudbath. There was no shelter. Their tenacity, belief and resilience in such conditions was astonishing, humbling and heroic; their tactics of resistance full of humour, humanity, patience and courage; their protest inspired by the Suffragettes and the spirit of underground Resistance in Europe during the Second World War. They built fires on the approach roads to the Common during exercises, preventing military personnel and occasionally their families and the fire brigade (my father included) getting in. Threats to shoot any incursions towards the inner fences (there were seven of those) were tested repeatedly as the women breached security to dance on top of the bunkers, in full view of armed American soldiers. Their bravery, energy, sense of community and form of non-violent direct action set the pattern for modern protest and inspired a generation.

The cruise missiles left the airbase in 1991, after the Cold War ended and the airbase was closed. The Ministry of Defence attempted to extinguish Commoners' Rights, the ancient liberties attached to general (usually unfenced) public open access and the rights of local dwellings to pasturage (for grazing animals), pannage (to let pigs forage in the woods), estovers (to take timber and firewood), and other uses of the land for household or individual use. Instead, they proposed a development of commerce, recreation and housing, while other plans threatened a 'third runway' for London, some 60 miles away. After much legal wrangling and passionate local argument, Newbury District Council announced plans to recover and restore the open Common. Meanwhile, a report by scientists based at Aldermaston's Atomic Weapons Research Establishment in 1961 was leaked by CND in 1996; the report had discovered unexpectedly high levels of radiation surrounding Greenham's runway in an egg-timer pattern. This was initially attributed to an otherwise undocumented incident of one or two fires on B-47 aircraft loaded with nuclear weapons in 1957 or 1958. Dissatisfied with military explanations and/or denial and pressed by local people, especially those directly affected by the leukaemia and cancer cluster, the council commissioned its own independent radiological survey, which was carried out by Southampton University that same year. Whether they had been before, no raised levels of uranium or plutonium were found.

On 24 March 1997, the Greenham Common Trust, a partnership of local authority and business ventures, bought the airbase for £7 million. The old airbase buildings on the southern edge of the Common became New Greenham Park, a business enterprise run on a not-for-profit basis, giving grants to local charities and community projects. The rest of the Common was sold to West Berkshire Council for a notional £1. On the open heathland itself, conservation was happening on a grand, ambitious scale as an enormous restoration project ensued. A million tonnes of concrete and asphalt were broken up and crushed, and decades worth of contaminants, pollutants and the leakage of huge amounts of fuel were cleaned up in a groundbreaking, environmentally sensitive process of bioremediation. The heathland was regenerated, pools were reinstated and created, surveys undertaken, and cattle and ponies introduced to graze between new cattle grids. In places, much of the military hardcore has been left as a memorial to what happened here. The bunkers are scheduled national monuments and the only part of the Common sold to private owners, the health and safety aspect of them too onerous for the council to commit to. At this time, aged 28 and working in the headquarters of the conservation charity BTCV (now TCV, or The Conservation Volunteers), I took the opportunity to volunteer with a group that created a bat hibernaculum out of an old command building – now inhabited by one of our largest, highest-flying bats, the noctule, as well as several generations of barn owl. The control tower with its fantastic panoramic view has become a wonderful little exhibition space and café, dedicated to the themes of the Common's military history, its history of protest and its wildlife.

Some 70 years after Charles Rothschild's pioneering, visionary plan for nature reserves declared Greenham 'ruined' by war, the movement his society grew into, The Wildlife Trusts, is continuing the work of West Berkshire Council to restore it. The Common has once again *become* common; common to all, shared ground, with all interests, from grazing rights to dog walkers, conservationists to historians, runners, riders, foragers and amblers, given free and open access. A common treasury.

After 19 astonishing years of continuous protest, the last Peace Camp at Greenham packed up in 2000 and, along with other local people and campaigners, I was there to finally walk back onto that open space and reclaim the land. But five years before we did, curving away from the western edge of my hometown, a new conflict was building.

CHAPTER THREE

Ten Thousand Trees

I was still waiting for the cuckoo on the Common when another bird came into my life. In my mid-twenties, I went back to full-time education to study English Literature at King Alfred's University in Winchester. After completing my 'A' Levels, I couldn't decide what to do next. Only one cousin had gone to university before in the whole family on both sides. I dreamt of applying for a fine-art degree course or one in English Literature but lacked confidence. I loved the idea of agricultural college, but wanted to study wildlife conservation. This didn't seem to exist on its own in 1988, but I came across it as an 'add-on' to 'Game and Wildlife Conservation Management'. I applied, and was warned there were no women on that course and was my career path to be a female gamekeeper? It was unheard of. While I dithered, a family contact of Dad's, who had emigrated to the Canadian Rockies and set up a horse ranch that took guided trips into the mountains, offered me a summer job. I'd never been on a plane, and had only ever left the country twice before, on school trips to France. I ended up staying for six months, sharing a remote log cabin in the Canadian wilderness with Michelle, another girl my age, and a young cowboy called Tom. I almost stayed.

Instead, in a time before social media, emails and the internet, and when communication was a letter sent by sea (airmail was too expensive), I came home to find my friends had moved on. I temped as an office assistant for as long as it took me to save up the airfare back and got a job as a cowgirl on a horse, then on a cattle ranch, for eight months in Alberta, with Tom and Michelle. I went to rodeos, took part in cattle roundups, had tea with Hutterite families (rural, religious communities of Anabaptists, similar to the Amish communities), camped out under the northern lights and broke in horses. Again, I almost stayed. I considered applying to agricultural college there, but it would have been hard to support myself and even harder for an English girl to get work on a farm afterwards in this very traditional rural backwater. I let my visa run out, and returned to England.

From there, I worked with horses, temped in offices, waitressed, worked as a chambermaid and as a shop assistant in several kinds of shops, all while living at home. I wrote poetry and painted landscapes and pictures of people's horses. I also attended as many evening classes in writing and art as I could. I read and read, and I hated working indoors. The dream of university grew stronger, despite my parents' bemusement and concern. No, I couldn't see a career path either. I just wanted to learn.

Reading English Literature at King Alfred's University was a hard-won dream, and for the first year I managed to combine it with Fine Art and Media Studies in a 'modular' degree. It was a pity it didn't also include a farming element. I loved walking around the city, through the cathedral grounds and out into the nearby countryside, through the water meadows of St Cross, or writing in a small café corner, nursing a pot of tea for as long as I could. I spent the week living in Winchester half a mile from where we'd lived as a family when my brother had been born, and not a quarter of a mile from Twyford Down.

Two years before I got the place at King Alfred's University, I was moved to join what was to become the biggest road protest in Europe, up to that point. The new M3 extension was almost built through Twyford Down, just south of the city. There was already a gaping chasm slicing through the oceanic flow of chalk hills, as if two strokes of a giant axe had cut it off from the old city and beheaded the high dome of St Catherine's Hill with its ancient turf maze. By the summer of 1992, the picnic-soft grassland of wild thyme and salad burnet, and its skylarks, corn buntings, yellowhammers and chalkhill blue butterflies, were already gone. With 5,000 other people, including my boyfriend Rob, whom I'd met at the record shop where we both worked, and our friend Ian 'Wob' Williams, a singer-songwriter for whom Rob played drums, I trespassed several feet below the height of where the old drover's tracks and holloways had been for centuries and what was now a hollowed-out chasm. Everything beneath our feet, around us and up to the horizon was blinding white 'high chalk' in its purest form. The air was thick with it, the sky infused with it. It felt like we stood on another planet altogether, on some bewildering, featureless moonscape. It was astonishing that something as solid, familiar and knowable as a hill could be so lowered, cleft through and changed. Its thin, green, living skin stripped away to leave a shocking, bloodless bone-whiteness beneath. But I knew I'd come in at the end of something, when all hope was lost. I'd seen and read the objections to the road in the papers; I'd watched the protests form and unfold on the news – and hadn't done anything about it except wring my hands. Any registering of protest now, while important, could only be symbolic. We formed a chain and solemnly passed lumps of loose chalk back up the hill, piling it into a cairn. I returned home in shock, unable to take in the scale of what I had witnessed.

I spent the next 18 months saving for my grand plan to go to university in Winchester, working weekends with horses and in various other full-time jobs during the week, all the while living with my rather long-suffering parents. Home was the last house in a medium-sized rural town. Yellowhammers sang their *'little bit of bread and no cheese'* song to one side of the garden and beyond, in an unbroken chain through fields, woodland edges and hedgerows, all the way to the near view of the yellow gorse on the downs. A little stream that acted as a storm drain for the houses on the estate ran past the bottom of the garden and through the roots of four great chestnut trees. In winter it flooded and in spring, the roots sprouted the strange saprophytic flowers of purple lousewort. Late that summer, Dad cut the hedge between the field and our garden, a wonderfully thick habitat of blackthorn and hawthorn, chestnut, holly, ash and oak, dogwood and dog rose. He cut it back hard, effectively coppicing it. I had become a walking 'Environmentally Sensitive Area' and was distraught. This thicket hedge was an entire ecosystem, a hidden world of thorns and birds and numerous voles, grass snakes and hedgehogs that now seemed unnecessarily and disastrously exposed. But the following spring, invigorated, it thickened anew and attracted the return of a visitor that had shunned its growing density as a former range grown too wild and woody for its exact purpose. Late April, a nightingale sang.

There was no mistaking it. I have never heard such song. And I knew those false nightingales. I knew that what you were most likely to hear singing at night was a 'most likely bird'. A big-eyed, light-drawing dusk singer, an all-night lamppost caroller – a robin, perhaps, a blackbird or song thrush. I knew their songs and this bird was like nothing else. It sang loudly, lyrically, with musical crescendos and dramatic pauses. It woke me in the night; it caused me to slow or hold my breath to listen. And when I did, I closed my eyes to let it in. It set a precedent that year by arriving on my birthday; a more exacting bird than the cuckoo.

Four springs in a row, it came. I would come home from university for it, as it seemed to come home to me. More often than not, it seemed to arrive in my dreams and immediately start singing (no time to lose) in the middle of the night, its loud song travelling through the window I'd left open for it, to penetrate and mingle with my subconscious. I was reading John Keats. I was living in his former haunt of Winchester. It was magical; it was haunting. It both soothed and nagged at me. I was on the cusp of making a new life for myself and trying to work out whether it was one my boyfriend Rob wanted too. I never really saw the nightingale, occasionally glimpsing a chestnut tail, although I hunted for it as if finding it would answer questions I could not even frame. It was a skulker in the shorn hedge. Just out of reach. A bird hidden in plain sight.

But then, in 1995, three years after our summer of trespass at Twyford Down, and one year into my university course, the Newbury Bypass, which until now had only been something to write impassioned letters about, became a reality. It was going to happen. Arguments had rumbled on and routes mooted for many years – it was rumoured initially that it was meant to provide ease of access between Greenham Common and the M4. The nine-mile route curved around the town and would smash through heathland, ancient oak and ash woods, water meadows and chalk streams, farm and downland. It made the M3 extension through Twyford Down look like a minor indiscretion. In its path lay 10,000 mature trees (one of which was reputedly the second tallest in Europe), four Sites of Special Scientific Interest (SSSIs), several local nature reserves, two Civil War battlefields, 12 archaeologically important sites, and National Trust land. And all of this within the North Wessex Downs Area of Outstanding Natural Beauty – just one rung down from a National Park. All channels of communication and all ways of stopping it were closed. In January 2006, *The Guardian* newspaper summed up what happened:

> *For the next three months, in scenes of high farce and near tragedy, a thousand people lived up trees, down holes or in rudimentary camps along the route while the British authorities tried as hard as they could to get them out...an army of several thousand police supervised 2,000 security guards, professional climbers, private detectives, bailiffs, sheriffs and others employed by the then Conservative government to evict them. The protesters went to extraordinary lengths to defend their trees and camps but when the battle formally ended on April 2, more than 1,000 people had been arrested at a cost of over £6m.*

To varying degrees, the protestors were joined by thousands more local people. The final cost of the protest alone was £24 million.

One night, just after Christmas, at our friend Ed's house in the village I now call home, five of us – Simon, Beth, Ed, Rob and I, who had all met through our shared taste in music at the Newbury record shop Rob and I had worked in – watched the first skirmishes unfold on the national Ten O'Clock News. It was a jolt, a clarion call to action. This was our home. Had been – and still was – our playground, and here we were, sat round watching it all fall on TV.

The following day, along the disused railway line, we stumbled into a battlefield. The old red-brick railway signal box that bore my name, written across its walls in marker pen by my teenage boyfriend almost a decade ago, had already gone. A man, perhaps in his early twenties, yelled at me to give him a leg-up into a tree. I hesitated, confused and frightened by what I saw around me. There was pushing and shoving and, moments later, the tree he'd been trying to get up was felled, horrifically close, the smell of pine resin exploding into the air. I leapt unthinkingly at someone wielding a chainsaw and felt two big sets of arms link round mine. To prevent me reaching the chainsaw, two security guards either side of me had co-opted me, unwillingly, as a struggling, furious part of a walking, human chain-link fence, escorting the chainsaw, helplessly and backwards, to the next tree.

Daily, for the next few months, the five of us woke and went from our homes on foot and on bike, meeting under significant trees or derelict railway bridges. There were careful, quiet dawns on the high ground of Wash Common, site of a bloody Civil War battle in 1643. We sat on the swings or stood on the tops of the neatly mown Bronze Age round barrows that doubled as memorials to the Civil War dead (and, as we'd believed and commemorated as schoolchildren, contained the dead soldiers and their horses). From this vantage point, as the Bronze Age communities, the Parliamentarians and Roundheads knew before, marching armies could be seen advancing. Almost the whole route was visible and we'd watch for distress flares from the camps or a slow convoy of juggernauts, police, horseboxes or portaloos nosing down impossibly narrow country lanes. We used a 'telephone tree' system that had first been used by the Peace Women at Greenham Common to alert fellow protestors. Each of us had the numbers of three fellow protestors that we would ring, often from the nearest public phone box, our pockets jingling with 20p pieces, should we see any action taking place. In turn, they would ring the three people on their list and the rallying call to action would spread silently, like the branches of a tree reaching out. Over such a vast swathe of countryside, the action erupted at different points, often at the same time. We went where we felt we were needed or to the places that meant most to us. Some of the flashpoints and battles became epic ones. Televised daily, they took on the names of some of the 33 protest camps: the Battle for Rickety Bridge, for Reddings Copse, Penwood or Granny Ash, for Middle Oak, Snelsmore, Mary Hare or Castle Wood.

And then there was the daily aftermath of a freshly broken, smashed countryside, scorched earth and bird's nests floating in the oiled eddies of the chalk stream when the contractors finished for the day and went home. We rallied and regrouped, took a moment with whomever we were with (local protestors, friends or those that had come in to help) to breathe and take stock of what had happened that day and to plan what we might be able to do the next day.

Rickety Bridge was a particularly special place, a wild, boggy, wooded island surrounded by the River Kennet, its tributaries and the canal. It was accessible only by a high, wobbly, single-plank bridge, without a handrail. The water babbled, gin-clear and deep for a chalk stream, under overhanging willow, where brown trout hid in the shadows. Reed warblers sang in the phragmites and flocks of siskins and redpolls twittered in the alders. There had been water voles here, but by now, this remote, all but inaccessible place was populated with machinery and a lot of shouting people. Someone had marked out the width of the road with the shock of a stuttering, white-paint line marker across the meadow. The place was no longer mine, but more mine than ever; it had been taken over by a protecting army and was under siege, its very existence at stake, from another. When we arrived, dozens of people were up in the trees, confidently but precariously balanced, some with climbing harnesses on, others suspended between two ropes strung between trees as walkways, or clinging, freestyle, to the very tops of the twigs, like birds. Many had padlocked their arms into plastic pipes, set with a metal rod into oil drums filled with concrete and hauled into trees. A sound system was strapped into the boughs, pumping out music, and the people in the trees beat up a rousing rhythm with pots, pans, whistles and drums. The emotion that the music inspired was something devastating, gut-punching through us with adrenaline and resolve; a call to arms taken up by the sound, always, of the poplar trees.

Penwood and The Chase were places of pine and cherry laurel; Snelsmore, oak. But here, down by the river, was a shivery, glassy place of poplar, aspen, willow and alder. Tall trees of a cool climate, poplars are watery in every sense. The hard, glossy, spade-shaped leaves have a serrated edge that catches the slightest rumour of a breeze, and their long, flat stalks cause them to wobble wildly against their neighbour. The trees interpret the breeze, giving it voice, and are never still, even when other species around them are utterly unmoved and silent. Hercules wore a sprig of fire-resistant poplar around his head as he descended into the fiery underworld. His sweat blanched the underside of the leaves, while heat blackened their surface. And Christ's cross was apparently made of poplar wood, the tree trembling and repeating its anguished whisper of grief ever after. Although dark green and silver-backed, the leaves appear black and white, flashing dramatically in strong winds.

On sunny days, the whispering peculiarity of the poplars added to the excitable tension of the day, a restless rustling, a beating up of courage. Yet at other times, they lent an atmosphere troubled with a quaking, nervy, melancholic sighing, the leaves thrilling to the sound of imagined rain. They created a shivery frisson that seemed to travel directly from the fine hairs on the leaves to the fine hairs on the back of my neck, reaching a hissing crescendo before the terrible roar of the chainsaws. The felling of the poplars called to mind poems I'd been studying at the time in an exploration of my final-year dissertation topic on an ecology of poetry and environmental protest. William Cowper's poem of 1784 bids 'farewell to the shade/And the whispering sound of the cool colonnade' of a much-loved stand of poplar trees, now felled, and reflects on the brevity of joy and, indeed, of life. Charlotte Mew's posthumously published 1929 poem 'The Trees Are Down' haunted me: 'But I, all day, I heard an angel crying: "Hurt not the trees."' Within days, these spectacular 'unresting castles' were, as Gerard Manley Hopkins lamented most poignantly in his 'Binsey Poplars, felled 1879': 'all felled…Not spared, not one'. They lay with their leafy tresses in the water, alongside neighbouring alders that, when felled, had the quirk of seeming to bleed redly into the water.

For three days, the five of us got onto the island and watched the convoy come, escorted by dozens of police riot vans, a horsebox and 16 coachloads of security guards. They were met by a crescendo of drumming from the trees. We acted as a vanguard, linking arms and forming a human barrier with other local people to block access to the trees from the ground, and were quickly, roughly, dragged off, the battle then taken to those in the trees who hung on with awe-inspiring tenacity and bravery. We sat in a line, then, on sodden ground, until the cold machinery pushed up against our backs and we felt the ground dip and ooze between our fingers. Police horses were forced between us and the heavy vehicles, and we were dragged from here too – some gently, some by a limb or even their hair. Nasty, misogynistic, sexist insults were rife, as they were at Greenham. The diggers and cherry pickers moved like medieval siege towers, wheeled catapults and *trebuchets* down the fields, escorted by a walking army in yellow to the besieged people on the bog island below. When the last of the security guards retreated and left each evening, the people in the trees abseiled down like spiders in the sunset. Accompanied by whoops, cheers and a victorious outpouring of eloquent and rousing prose, it was like watching some revolutionary theatre production.

At Reddings Copse, the eviction was halted in a week-long stand-off. A specialist hydraulic platform, a monstrous cherry picker from Holland (reputedly the largest in Europe), had been hired to oust the highest tree house in one of the tallest trees in Europe. The Corsican pine, visible from most of the undulating route, had a 15-foot circumference at its base and was 150 feet tall. It had no branches for the first 75 feet and, in true outlaw style, a longbow was employed to fire a line over its lower limbs in order to scale it and start construction. The tree houses – for there were two – contained a kitchen and a wood-burning stove. A local man, Balin, who had spent 16 days in lone vigil at the top of a tripod on the route, free-climbed his way to the besieged tree through the canopies of neighbouring oaks, 80 feet high, which had friendlier branches. When the cherry picker was extended to its full, rearing height of 200 feet, it seemed to hesitate, palling as the protestors unclipped themselves from safety lines in audacious, heartstopping response. But instead, just as the Roundheads and Cavaliers had done in this very spot in 1643 and 1644, the bailiffs in the platform went for the opposition's colours, laying aside bolt cutters and chainsaws and leaning over to pluck the flag that fluttered from a ladder extending from the tip of the great pine. As the colossal neck of the machine bent to lower the flag to the ground amid jeers and cheers from both sides, an oak was pushed over simultaneously by a digger. It twisted awkwardly and fell with a screaming render of wood onto the body of the cherry picker, damaging it irrevocably before bouncing off and falling on one of the bailiff's climbers. He was knocked unconscious and taken to hospital. The Under-Sheriff of Newbury, ever present in his spotless red coat, continued to eat his lunch. Many protestors were injured during the campaign, as were a few bailiffs and security personnel. There was a dangerous, cavalier disregard for safety.

I'd neither the bravery nor the head for heights for such bold action but, nevertheless, I learnt new skills and a new confidence. I walked tightrope walkways between banks and islands, and between trees; I tied knots in pipes pumping water into run-off pools (once, with my mum); I flicked cable ties into chainsaws to stall and break them; I wept, charmed, persuaded and railed against security guards and into the video cameras of the silently sinister Bray's Detective Agency (a private investigative company contracted by the government in an unprecedented move to undertake surveillance of any individual attending the anti-roads protests of the

1990s). I took food up to those in permanent camps, fell out of trees and into rivers, canvassed, wrote letters, sat in police vans listening to a sympathetic telling off by local policemen and took the fight into town, too. Particularly as local people, we were determined that the strength of our feelings would not bypass the town as the road would, and that we would make our small mark on history beside the bigger one the Greenham Peace Women had left. Once again, the town was divided and angry. As at Greenham Common, most pubs, restaurants and sometimes shops banned the 'protestors'. The local newspaper, the *Newbury Weekly News*, tried to keep its editorial neutral, with debate raging in its letters page, and it seemed in every group of people – whether queuing at a till or at the post office, people had an opinion. Once again, my friends and I seemed to be in the passionate minority.

Back in Winchester, I'd retreat for two or three days to go to lectures on the Romantic Poets and salve the hurt. But there was no hiding. Their words resonated and became like the drumbeats of a call to arms. From Blake and Shelley's radical voices, to Wordsworth and Coleridge's protests about railway branch lines reaching into the Lakes and, in particular, their idea of nature being beyond ownership. In his *Guide to the Lakes*, William Wordsworth said that nature should be 'a sort of national property, in which every man has a right and an interest who has an eye to perceive and a heart to enjoy'. Even in Winchester, I found veterans of the 1992 protest against the M3 extension through Winchester's Twyford Down, and we formed a 'Support Newbury' Group. We intercepted convoys of security guards coming up to Newbury through the freshly cleft landscape and learnt to check meeting rooms for bugs. Once, provoked by a shopper's comment on a tabloid newspaper headline about dirty protestors on benefits ruining life for local people at the tills in Winchester's Sainsbury's, I got up, front of store, and delivered a lengthy, emotional riposte. I'd worked *hard* to get where I was. I was not claiming benefits and I was local. Not only that, I also didn't want an ecologically destructive road running through my home, adding to the environmental crises, and would be forever grateful to those protestors who were defending it, living rough, when I couldn't be there. I earned a round of applause from the tea and coffee aisle.

I learnt a lot about people. There were cliques and mistrust and acts of life-affirming solidarity; brutality, and sudden kindness from strangers. I got away with a lot, mostly, I think, because I didn't dress like a typical protestor – I often wore a skirt (that I could run or climb in) with a cardigan and a pair of brown boots. I was never hit or deliberately hurt myself, but I saw it happen to others.

I was verbally abused, shoved and dragged off. I learnt never to judge and that nothing is ever entirely what it seems. Security guards defected to the other side – indeed, the last man down from the trees at a later road protest, at Fairmile in Devon, was an ex-Newbury security guard – although nobody went the other way unless you count, temporarily, *The Guardian*'s Environmental Editor at the time, John Vidal. He infiltrated Reliance Security, the firm employed to deal with the protestors, and exposed a culture of violence.

On a long stand off on a gloomy day, when all we could do was to set ourselves facing the guards on the other side of the cordon, a flock of yellowhammers and chaffinches came down to feed on spilt grain along the farm track. 'Canaries!' cried one of the guards. 'Escaped budgies!' I laughed and told him that these were indeed our own bright canaries, but that they were true wild natives – and were in steep decline. For a good 15 minutes, the streaky gold birds with their September wheat-stubble colours drew an appreciative audience from both sides. The security guards marvelled at such pretty exotics in the damp English fields, alongside such clear water. How beautiful it all was here, they said, how peaceful, how lucky you are and how hard this must be. They shook their heads and discussed how to get hold of birdseed for the yellowhammers.

Some of the police were locals trying to remain neutral, and walked here in the countryside with their families at weekends, between shifts. Among the security guards were ex-prisoners and curious students, family men forced away from their home and living in miserable conditions, just to send a wage back. One was a former miner who had been on the picket lines throughout the Miners' Strikes. When a close friend, who had supported him through those unimaginably hard times, died, he could not afford to go home to the funeral. There was the blue-dreadlocked, pierced man who came down from his tree to hug me as I wept when one of the 500-year-old oaks was felled. He shimmied back up his tree, ensuring a reprieve for it for two more days. On another occasion, a man who wore a skirt and a lost, kindly expression came crying along the sunken road after a raid that took us all by surprise. 'I found something to believe in!' he wept. 'I only went to make a phone call and they took it down. I've been here four months!' All I could do was hug him. These were humbling experiences.

———

Towards the end of the Battle for Rickety Bridge, a kind of desperation set in and actions became increasingly symbolic. A few people had dug up small trees and saplings on the route to replant them on the other side of the painted white line. After staying the night at my friend Ed's house, he and I arrived early, crossing the eponymous bridge with practised ease on what was a misty, late spring morning. The mist hung above the river and was shifty. A rising wind stirred the tops of the poplars so that they whisked the mist around like egg white. What I took to be a crab apple tree appeared in front of us, glowing with the brightness of its small, golden-yellow fruits. Without a word, we moved towards it, when the apples exploded into life – a dozen yellowhammers that flew away twittering, lightbulb heads illuminating the mist like fireflies. I was bewildered, entranced; it hadn't occurred to me that this 'vision' of golden apples was entirely out of kilter with the season. Nothing was said, but we both knelt in the mud to dig the tree out with our hands, to save it. Ed, who seemed to share the ferocity, passion and conviction of my feelings and actions towards what was happening at the bypass, realised first and stopped digging, wordlessly. The tree was rootless. It had already been felled. Someone, from one side or the other, cynically or out of defiance or respect, had stuck it back in the earth. We both sat back on our heels, and I wept hopelessly for all the trees.

For all the battle talk, anger, grief and loss, we protestors weren't fighting for life or livelihood directly, but we were fighting for what we loved and believed in. It was a battle in the war on the environment, on place, landscape, wildlife, locality; the making way for the unsustainable juggernaut of future congestion. We were fighting for home in a small local sense, but also in a global sense, for the future of our only home planet. This was our Spanish Civil War, where no blood was shed, but where our ragged army was made up of a media-savvy bunch of poets, writers, intellectuals, lost souls, misfits and 'greens'. I'd watched and been moved by Ken

Loach's powerful film on Spain, *Land and Freedom* (1995), where the wrong side had won, of course – but, for once, history had been written by the losing side, and eloquently so.

There were small victories. One particular oak tree stands defiant today because someone had the foresight to map it and reason that it could form a traffic island. It was also the only tree blessed and protected in a Druid ceremony. After a long standoff, Sheriff Blandy met with protestors who agreed to come down from the tree if it could be saved. A tense treaty was drawn up and signed by both sides in front of the gathered media, but the brief truce ended when, as an ITN news journalist reported, 'the sheriff was quite literally run out of town' by outlaws hurling tomatoes. There was a lot of wry humour. There were the inventive and hilarious tactics of pantomime cows, a replica Spitfire 'crashed' into a tree, organised army-style manoeuvres designed to mesmerise and entertain the guard line before, on a signal, it was breached – much like when I have watched a stoat 'hypnotise' a rabbit with a manic display of acrobatics before pouncing.

My senses became primed: raw and heightened like never before. Sights, sounds, smells, the way everything felt, all took on a new, acute significance. The burning of each wood smelt different – the sweet resin of pine, the fireside warmth of ash, the toxic, dense blackness of burning rhododendron that provoked a frightening cough which is still often with me today. I was changed. I looked at the world differently, felt it more keenly, knew it better, was more alive than I'd ever been. And all that spring, back home and surrounded by the whine of chainsaws and the constant plumes of acrid smoke not half a mile away, the nightingale sang.

CHAPTER FOUR

An Extinction of Nightingales

This middle-class, middle-England rural town was no stranger to protest and anti-establishment views. There is a timeline of resistance recalled on the walls of pubs and coffee shops, and remembered in the names of beers brewed locally. Whether it was the freedom to express religious or political will against those in power or the fight for the right over access to the land, a home and a living wage, there has long been a seam of rebellion in Newbury. Even as you approach the town from the north, there is a roundabout and gyratory system called the 'Robin Hood Roundabout' at the beginning of what was the first Newbury bypass. Named after the Robin Hood pub beside it, established in 1796 and now a Toby Carvery, the history behind its name is much older. Though the facts behind the legend of the outlaw Robin Hood are notoriously various and hard to corroborate, the second oldest mention of such a man defying authority dates from 1262, of a man from Enborne, near Reddings Copse, on the route of the bypass. 'William Robehod, member of a band of outlaws' is referenced on the 'fine rolls' of King Henry III, which were legal documents written in Latin on rolls of parchment by the royal chancery, detailing (among other things) the seizure of property for, in this case, unknown indictments.

Newbury had its martyrs during the Wars of the Roses as well as those who challenged the Reformation in 1556. Julins Palmer, scholar and principled teacher, was tried and found guilty of asserting his religious belief, of sedition and religious heresy. He was publicly burnt along with two others, John Gwyn and Thomas Robyns, on the outskirts of town, somewhere near the bypass route. Palmer had suffered as a Roman Catholic under Edward VI and lost his teaching position at Oxford. After being reinstated during Mary's reign, he became powerfully affected after witnessing the public burnings of Protestants, experiencing a dramatic turnaround in his belief, renouncing his old religion and converting to Protestantism. He suffered during Mary's reign as he had under Edward's, was betrayed by friends and disowned by his own mother. Again, he had no home or means. Yet Palmer spoke so passionately, eloquently and with such conviction that two of the jury, the local Justice for Peace and the Sheriff, offered him support, a living, land and a way out, if he would only repent and agree to live quietly. But Julins Palmer would not repent. Among his last words spoken were: 'We shall not end our lives in the fire, but make a change for a better life.'

Over the same ground and almost 100 years later, two key Civil War battles took place. Some of the 600-year-old major oaks, felled in minutes, had sheltered Cromwellian and Royalist soldiers. Newbury had Parliamentary sympathies. In 1643 and then again in 1644, the fields and lanes on the bypass route were the site of desperate bloody battles. Roads and the pubs around the high point of Wash Common bear battle-scarred names in commemoration, but more telling, perhaps, are the local names that persist without being recorded on maps or satellite navigation systems: Battle Hill and Red Hill (because its camber and gullies ran with blood like heavy rainwater after a storm). Things like that are not forgotten. Neither are the ghost stories that hang around like wisps of mist, repeated and reaffirmed around the old farmhouses that received the dying and the horribly wounded. After I'd come back from Canada and before university, I worked as a groom for a time and rode horses that spooked or refused to go past the gates of Wheatlands Farm, Biggs Hill Cottage, the spectre of the old Cope Hall (demolished in the 1960s) and the renamed Falkland Garth. Viscount Lucius Cary Falkland, the King's Secretary of State and a peaceable, literary man, died there. He was fatally wounded after finding a gap in the otherwise impenetrable blackthorn hedge and spurring his horse through to ride at the opposing army alone. Witnesses thought his action suicidal; a man reputed to have a 'sweet reasonableness', he was utterly distraught that his own countrymen were killing each other. The 1st Earl of Carnarvon, Robert Dormer, was also killed on the same day.

Fighting for the Parliamentarians, on the other side, and on the same blooded field, was family man and cheesemonger Thomas Prince. Badly injured, Prince survived and went on to become one of the leaders of the Levellers. A progressive political movement, along with abolition of the Monarchy and the House of Lords, it eloquently and calmly demanded annual elections, religious freedom and the end of censorship. It called for 'just laws, not destructive to the people's well-being', trial by jury and tax breaks for those earning the lowest wages. In 'An Agreement of the People', with a clear-sightedness ahead of its time, the Levellers also advocated that those paid a wage, as 'servants of the rich', should not be allowed to vote, as they would be unduly pressured (directly or indirectly) by their paymasters. Once, during election time, on the doorstep of the cottage my husband Martin and I now rent, I was berated by an election-party candidate for displaying the two opposition party posters in our window. I was told, half-jokingly, that I should be ashamed, bearing in mind it was a Tory MP who had not only *built* our house, and whose son owned the country estate we lived on, but provided us with a home we may not otherwise have, and probably a job too (they assumed we worked on the Estate) and, they observed, fitted nice new windows for us.

Much stays the same. The battlefield hedges are still there, still impenetrable (though Falkland's mortal gap has long closed), and history does not record, of course, the stories of the common foot soldiers, musketmen and pikemen. But in the museum in Newbury, part of the Civil War collection includes cannon and musket balls turned up in the garden of my parents' 1970s bungalow. In September 1878, near the spot Viscount Falkland was killed, a plain granite obelisk was erected, on the former (now enclosed) common ground. But it was not a monument to all who fell; it was a monument in memoriam of Viscount Falkland, unveiled and dedicated by the 4th Earl of Carnarvon, whose predecessor had also been killed on that day. The High Sheriff of the 'Royal County of Berkshire' was present. The decision to take such a strong, Royalist angle on a Civil War battlefield memorial is still controversial today. The inscription on the memorial, indecipherable to most, is entirely in Greek and Latin.

Here too, at hawthorn-blossom time in the spring of 1795, the precursor of the Welfare System was created in the

An Extinction of Nightingales

hamlet of Speenhamland to provide relief to the rural poor, after nationwide Bread Riots. At the Pelican Inn, a minimum income for agricultural workers was established and linked to how many children the labourer had and the price of bread. A scale of top-up credits, coming out of rates, and similar to many modern tax and welfare systems was agreed. The Speenhamland System (or Berkshire Bread Act) became the best known of several similar schemes and 'bread scales' in England. Yet it did not prevent some of the most violent activism, rick-burning and machine-breaking during the Swing Riots in the 1830s among local farms and estates. The Speenhamland System was abolished by the 1834 Poor Law Amendment Act, which ushered in the workhouse system. At secondary school in the 1980s, the History curriculum included the Industrial and Agrarian revolutions. We learnt how many local rural families were sent to workhouses when their breadwinning men were deported for protesting: for making a united stand against the combined injustices of falling wages, tithes and new, labour-displacing machinery. The Justices of the Peace and the Sheriff of Berkshire called them to account, without much sympathy. Those same titles had done for Julins Palmer in the end, and oversaw the daily evictions on the bypass.

We didn't know it at the time, but, 150 years on from those desperate riots, the names of several families had been blanked out on our photocopied sheets, lest anyone realise the workhouse families shared the surnames of classmates now in receipt of free school meals. One, a William Winterbourne, was sentenced to death for his part in the riots and pardoned in a local ceremony only recently. He was a relative of the tractor driver who works on the farm estate where I live now.

I grew up knowing all of these narratives of place. The community retelling them in the form of ghost stories, or a series of outdoor, processional plays or re-enactments by the Sealed Knot Society. Our local police constable, PC Stubbs, took great pains to tell and dramatise the histories on landmarked walks, where the Bronze Age (and earlier) burial mounds or 'barrows' doubled up as the turfed-over, mounded graves to the fallen men and their horses of the Civil War. When attempts to level 'the bumpers' were made in the mid-19th century, human remains presumed to be from the Civil War were unearthed, along with buckles, bullets, bits and cannonballs, from 1643 – and the sounds of phantom battles. During one bypass battle standoff, in the mist, near the site of Lord Falkland's suicidal plunge, two loose horses appeared from nowhere, pushing through a gap in the hedge, and proceeded to gallop among the security men and protestors, squealing and snorting at the police horses. There were several accounts too, from security guards, of Roundhead soldiers marching through the river valley with pikes on their shoulders. Just as I had done at Greenham Common, we stirred the sense of unease, asking security guards if they were on duty at night in the valleys – did they see ghostly figures marching? One I spoke to refused to go on watch near the river at night.

I grew up needing to belong to the landscape's narrative: for me, the buildings, trees and trackways had retained, absorbed and exhaled their old stories and somehow renewed and retold them. The evidence of its peopled history was there in the annual laying of flowers on the memorial mounds, in the healed hole of the impenetrable blackthorn hedge, in the hayfields, and in the way one of the mounds, recognised at last for what it was, a round barrow or tumulus now cleared of encroaching scrub, threw up a thatch of bell heather, ling, furze and petty whin. The heathland had resurged with a flammable, spiteful fistful of remembrance that recalled its Common usage for thousands of years: heather for stuffing mattresses, birch for besom brooms and prickly armfuls of 'fuzzy baggins' to fire the bread ovens with, when there was any bread to be had. Most of the Common was taken and parcelled up in an act of enclosure in 1855, by those who already owned land. Perhaps in the outskirts of any small market town there is a history of battle; a necessary seam of uprising and dissent, just as there is use and abuse of power, wealth and standing. Nevertheless, we, the protestors and the remnant rural poor, finally felt that here we had history on our side. That this theme of resistance, and of not accepting an authority that did not speak for us, had grown beyond the right to small parcels of subsistence land, or a living; it had grown to encompass the planet itself. History would surely show that protecting the air we *all* breathed and the earth we *all* stood on, and the sharing of finite 'resources', was no longer about 'us'; it was about life on earth itself and the right not to have it eradicated through nuclear warfare or ecological disaster.

Back on the bypass, I'd fling open my bedroom window and listen to the nightingale. Trying to describe a nightingale's song does not convey the impact it has on the senses, the emotions or even the nerves, especially when all those things are heightened to such a pitch and primed to receive it. The poet John Clare described it as an 'out-sobbing'. A series of long, high whistles on the same note would quietly needle into the subconscious first before building to a crescendo and dropping to a low '*jug, jug, jug*' beneath the sweet melodic chorus. The volume is astonishing, tropical, and particularly amplified at night, when there is little noise and no other birds singing. The pathos of its carrying power, depth and range of notes is breathtaking. It pierces the heart like Cupid's arrow. It moves you and pins you to the spot. A nightingale will sing hidden in deep cover, pausing theatrically, listening for a response and often utilising natural amphitheatres, particularly near water. There is something of the feverish, lovesick insomniac about its wildly exhilarating, aching musicality, sung with such precise, measured vocal power almost all day and night over the course of about five to six weeks, as if it simply cannot be kept up, sustained or endured forever. Perhaps that is why the singing season is so short.

Then, on those nights listening to the nightingale, I was a 26-year-old with few commitments. I soaked up the quiet, reflective atmosphere of the ancient City of Winchester when I could. I loved its water meadows, chalk streams and downland, all within walking distance of the city centre. I loved the cool Great Hall, the Cathedral and King Alfred's University buildings that mixed flint, gargoyles, steel and glass, that were set steeply into a chalk hill with views of the Matterley Bowl and Cheesefoot Head. I loved the fact that I could walk home, hugging my books to my chest, past Jane Austen's house, past where John Keats had written his 'Ode to Autumn', and that I might happen upon an otter in the old City Mill. One of my English Literature tutors, Howard Carter, reassured me quietly that I'd learn more about poetry during a revolution than in the classroom. I skipped a few lectures, and he was right. Relationships changed and were intensified for me by a political and environmental awakening and its subsequent fury. I loved and lost and loved again.

An Extinction of Nightingales

I once heard someone say on Newbury's main Northbrook Street: 'I couldn't be that passionate about something' and thought that was a life not lived.

In this landscape, every word, every nuanced smile, every eloquent speech or argument, every leap of faith and every daring action was played out against a backdrop lit by a mess of love and a countryside – my countryside – on fire. When the work stopped at weekends, the five of us took a break and ran to the hill that formed the distant view from my bedroom window, over the head of the singing nightingale, to gain some perspective. Walbury Hill is the highest chalk hill in England and on its summit – it is, in a modest, South Country way, just a few feet short of a mountain – are the ramparts of an expansive Iron Age hillfort. From here, the curve of the bypass was visible as a smoking wound illuminated by flashing orange lights in the valley below. This was and still is a place of dark night skies, solace and silence. We sat on the grassy ramparts, among the harebells, watching the bright comet Hyakutake appear, its 360-million-mile-long tail of ice, gas and dust streaming like a portent overhead. Choked with so much emotion, the chill of a clear night, the effects of inhaling rhododendron smoke and bottling up unspoken feelings, I could not stop coughing.

I'd sleep fitfully, nightingale song coming in through the open window, assailing me the minute I got home, often long after dark, and not letting up when I woke in the morning, sometimes to go out before dawn. When I wasn't there, I would angle my daily walk to lectures in Keats' footsteps, through the water meadows to St Cross. I thought of Keats' biblical Ruth, homesick among alien cornfields and brought to tears of home by a nightingale's song. The intensity of his nightingale's song cast a spell that charmed windows open and put the fevered listener, a young, dying man, in a trance between sleep and wakefulness. I'd fling open the window at home and listen. The music, for it was that, took on a life of its own separate from the bird itself; its meaning, though it emanated from a bird hidden in a hedge, was destined for interpretation by whomever heard it. Sometimes I'd close the window hard and curl up in my single attic room bed, my face to the wall. But I could not shut it out. All the while it was urging me: 'Live, live, live, do something about this!' Sometimes, I wasn't even sure what 'this' should be. I could only listen.

Once, at Rack Marsh, near the tiny Watermill Theatre, whose director was a powerful voice and campaigner against the road, I heard a different nightingale sing. There had been a pause in the action, a standoff; the protestors I was with that day were cold and restless, rain pattering down onto the giant paddles of butterbur leaves as if they were hollow wood. Some of the leaves had been plucked for use as umbrellas. The pristine River Lambourn poured under the thin footbridge lined with hemlock water dropwort. On the hill above were the Civil War ruins of Donnington Castle. Not for the first time that spring, we stood, miserable and near-defeated, shuffling our feet where 17th-century troops had rallied, rested or despaired, aware of a certain irony, an incongruity, and, I suppose, embarrassed unease at the comparisons we were making. Yet beyond the castle ruins were the black silhouettes of the trees at Castle Wood camp, their limbs sawn off and blunted as if shelled, polyprop walkway ropes still connecting them, marooned in churned mud. One of the camps in these trees was made from a lifesize replica Spitfire. It made all the papers.

We were disconsolately searching the riverbank for the tiny Desmoulin's whorl snail, whose rare presence we believed might stop the road in the ultimate David and Goliath battle, when there was a sudden, brief, astonishingly loud explosion of song. A nightingale was skulking low among the tangle of waterside roots. A few people heard it, fewer knew what it was. I willed it, like the bird at home, to sing its heart out for all to hear, to put on a grand performance outside this famous little theatre – that would surely then stop the progress of the road forever. But it was petulant, reticent and didn't oblige.

Meanwhile, my bird at home sang longer and louder than any nightingale I've heard since. I think partly this was a physical need to be heard above the chainsaw and heavy machinery noise and the burning around it. Partly, because my senses and emotions were so heightened to receive it, but mostly, I realise now, that this was a lone, pioneer bird, reclaiming former ground held by its ancestors. It was taking a risk. It had a perfect territory with no competition, but as yet, no mate. For its own continuation and genetic survival, it had to sing a mate down from the skies or attract one from nearby, but there was no 'nearby'. The nearest nightingale population was a small one at Greenham Common four miles away. This was a virtuoso performance for nobody's benefit, a song in the dark of increasing desperation, power and beauty. Every note was precisely executed. The whistled notes that built to a crescendo in the classic *'shree, shreee, shreeee, jug, jug, jug'* were sung hard, drawn out slowly and carefully enunciated. It was a song of yearning. No nightingale I've ever heard since can hold a candle to this torch singer. Other nightingale song has been beautiful, instantly recognisable, but falls short somehow, the notes rushed, imprecise. Years later, the exact timbre, tempo and power of that little bird's song stays with me and it is this memory, fresh and powerful, that has become the benchmark by which I compare all other nightingales. 'Thou wast not born for death, immortal Bird,' wrote Keats. Indeed. I stop to listen and am always moved. But I cannot find its match. That the singer did not live to breed and pass on its singer's genes was heartbreaking. It should have been a survivor, an evolver, only the human race had shattered and isolated what it needed to do this. It was a loss to a world that never knew its loss.

With the clearance work on the bypass over, we five left Newbury to spend a somnambulant week in a complete, undamaged, Welsh countryside, finding beautiful, relatively undamaged places, skimming stones over oxbow lakes, camping out in a remote farm cottage that had been taken over by sheep. In a secondhand bookshop, I came across a collection of Australian poet Oodgeroo's work, *My People* (published in 1970). Lines from her environmental poem 'Time Is Running Out', urging action for the planet, strong as love to match those that sought to destroy it, struck a chord. We crossed and recrossed mountain rivers, leaping over treacherous rocks, and explored the profound peace of the coal-black velvet darkness of old railway tunnels. And over a reservoir, I saw my first red kite, circling above its reflection in the glassy water, red, black and white feathers elegantly ragged, like a Native American's headdress. It comforted and thrilled me. Surely, if we could bring these magnificent birds back, we were capable of good things? A dialogue with the wild world, perhaps, wasn't over.

Back in Newbury, I was home in a place I no longer recognised. What does a 'sense of place' mean when that place is altered beyond recognition, literally scraped off the earth? When you get lost in the place you grew up in? I came up against fences as I had as a child, when the bridleways of Greenham were suddenly dead-ended and denied with chain-link fencing and concrete posts.

On the 11 January 1997, seven months later and in thick fog, we had our last hurrah. The whole nine-mile route was now encircled by a razor-wire-topped fence, ten feet high. An event had been organised to mark the anniversary of the clearance work starting. It had been inspired by a visit to the bypass from former Greenham Common women along the lines of the 'Embrace the Base' demonstration that took place at Greenham Common in December 1982. The atmosphere was lively but good-natured and we moved to encircle the fence of the big compound which was now a great hole in the ground, full of machinery, an enormous crane and hundreds of security guards. As we moved to link hands, a mounted policeman forced his horse between us and the fence at a canter. Not everyone, including me, could get out of the way and I was knocked to the ground along with others. I wasn't hurt, but it was too much for the man I'd brought with me. A friend, whom I'd met at the stable yard where I still worked part-time at the weekends, had been until now a little cynical of the whole thing. He grabbed the horse's bridle, pulled the big grey horse up and remonstrated with its rider, narrowly missing being both trampled and arrested. In the mounted policeman's wake, the fence was rocked, Greenham-style, until it went over, and everyone poured into the compound. All around us, people grabbed the hats of the various ranks of security guards, red, yellow and white – some were given up willingly, the atmosphere still light, if a little devilish – and began drumming up a rhythm, banging the hats together and against the diggers, crane and dumper trucks. There was a carnival atmosphere and little resistance; some of the guards clapped along with the general drumming and there was only the occasional sudden rush forward as rumours of a police charge rippled through.

Several people climbed, freestyle, to the top of the enormous crane, raising a flag and rousing the general music by banging safety helmets on the neck of the great crane – a huge cheer went up – and then everything caught fire. It was hard to see who, but someone, or several people, set fire to the heavy machinery, Land Rovers and Portakabins amid much shouting and recriminations. Tyres were slashed. The smoke and fog obscured much of what was going on and I shall never forget the strange and eerie images from that day: all that machinery burning in a huge chalk pit with someone riding the neck of the great crane as if it were an ancient fire-breathing dragon, flickering flames reflecting off the fog in a mockery of the construction vehicles' orange flashing lights. It felt dangerous, it felt justified, it felt wrong, all at once.

Afterwards, Martin, my rescuer that day, accompanied me whenever I felt the need to make a small, but seemingly pointless protest, since all the trees had now been felled and the earth churned up ready for the last phases of construction. That day in the fog, his eyes had been opened – and, it seems, so had mine to him. We made gentle attempts at mischief along the route, cycling along slowly, our bikes drawing together, hands and handlebars clashing. We trespassed and consoled ourselves by exploiting lapses in security, racing through open gates and once, driving his little Bedford rascal van, hazard lights on, up the road itself for several hundred yards before being chased back out. That spring, I waited for the nightingale and it never came. It never came again.

An Extinction of Nightingales

We went together, up the hill to watch a different comet in the sky. And this time, I felt that I could breathe. We lay against the encircling arm of the old hillfort's ramparts and turned away from the view of Newbury's distant lights towards the downs, softly folding into utter darkness. The comet Hale Bopp burnt above, its blue gas tail pointing away from the sun with its golden trail of ice and dust.

One night I walked Martin out of town and along the river and canal to discover the towpath blocked and a great, incomprehensible chasm open in the ground. I didn't know what I was looking at; neither my eyes nor brain could make any sense of it. There was no water where there should have been water, but no bottom to the chasm either; there were concrete sides into nothing, a great slide into hell. I felt cut off from everything I knew, surrounded by an unfathomable blackness, unable to get home. I was confused, emotional, and it frightened me. I hurled angry stones at the security lights with a raw, screaming rage, but they simply bounced off the reinforced plastic. Up to that point, I think I'd really believed it wasn't going to be built, that we could stop it. I clung to him, then, this man, for it felt that the whole world had shifted and tipped and if I didn't, I'd be lost, too. I clung on because I was falling, sobbing into oblivion. He held on. In the minutes that passed, beyond the reach of the lights, we watched the first of that year's Perseid meteors shoot across the sky.

A little of that evening has never left me. My mind sometimes has a blank spot, an unconscious forgetting or a denial that the road exists. Momentarily, my brain does not take in what it can see, cannot process tail lights that fly through the air on an elevated stretch of the renamed A34, like tracer curving above hedgerows. I walk footpaths that, in my mind, still cross hayfields; bridlepaths that skirt, uninterrupted, the heathland; search for the peace of the water meadows; and cannot process, for an unsettling heartbeat, the chain-link fence, the concrete bridge, the motorway embankment. It is like a shard of glass gone from a mirror, the piece missing from the eye before a migraine surfaces. The ground beneath my feet feels uncertain and unreliable. I need Martin then, his steadying, cheery and reassuring presence. Every April or May, I'll drive down to the bottom of the little cul-de-sac where I lived with my parents, and do a slow three-point turn with the window down. If no one is about, I turn off the engine for a while and listen for the nightingale. I've not heard it since. I'd witnessed an extinction of nightingales. That summer was the last for me there. I moved on too, renting a little house made of horsehair, chalk and flint, which was so damp that a builder declared the nearby River Test must rise from its very floor and up the walls, with Martin, the man that saved me and stopped the policeman's big white horse from trampling me.

CHAPTER FIVE

The Thing with Feathers: Fieldfaring

What kept coming back to me in those dust-settling days, weeks, even months afterwards was the vivid memory of the red kite's slow, mesmeric circles above its reflection in the Welsh reservoir. It crept into my dreams, circling on the periphery like something tethered to me. Something to hang on to. And during waking hours, behind my eyelids, I could picture with clarity fox-red primary feather-fingers, an orange, delta-shaped tail, black wrists, white flashes, a pale head. The recognition of seeing a bird back from the brink of extinction like that, unexpectedly, yet absolutely in the right place, set such a life-affirming thrill through me, it seemed at once to send my heart soaring and me back down firmly on the earth at a time when I badly needed it. It was something jubilant. As poet Emily Dickinson wrote: 'Hope is the thing with feathers', after all.

In the aftermath of the bypass, I was left reeling: emotionally bruised, upset, angry, lost – and also disenfranchised. I'd watched the place I knew and loved be utterly destroyed. Ultimately, there was nothing I could do about it. It was not mine in any tangible sense and I could not save it. I was barely able to comprehend the strength of feeling that assailed me. I was grieving. I hung on tightly to the image of the circling red kite and for a while, at times when I felt like crying, took to closing my eyes, picturing it and tracing its slow, soothing revolutions with my thumb on the inside of my wrist, for comfort.

I began to take stock, mentally, of what had been lost and what could be lost, wherever I went. I started to look closely at things again, in the way I had last done probably when I was a child or perhaps when I used to paint and draw in my late teens and early twenties, searching harder for the essence of things. I seemed to see all wild things and places through a 'last chance to see' filter. I began to wonder just how well I knew my own wild familiars? I determined to re-engage and set myself challenges; could I really tell oak from ash in winter? Fallow deer prints from sheep? Fox earth from badger sett? I knew how the different summer grasses felt between my fingers, but I did not know all their names. I decided that, for a long time, my relationship with nature had been passive; that deep knowledge was going to be my best defence and gaining it would help unlock the access I had lost. I felt, with growing conviction, that knowing a place and all that was in it, and how it related to me and the wider world, was as potent a key as any. It was a way back in, an entry point. A key to a place I could inhabit and know better than those who might seek to destroy it. It was a way to own it if I went back to the start.

I went back, in fact, to an afternoon on a bus a few years before the bypass, a few years before university. On my day off from my job as a groom, I'd soaked and scrubbed the ingrained map of dirt from my hands with lemon juice, put on a skirt and ridden the bus to Winchester. I loved the meander of the journey and used the time to try and work out where I was going. I thought seriously about applying for one of the new Wildlife, Game and Conservation Courses at the agricultural college en route and dreamt of studying English Literature or Fine Art in Winchester. I felt torn between the apparently incompatible things I wanted to do. I was completely unable to make a decision between them.

But my pivotal memory of this time is this: of a young woman travelling on the bus, a little older than me and sophisticated. She wore a tweed 1940s-style skirt suit and smart court shoes I much admired and, out in open country somewhere near Andover, she leaned across her companion excitedly, pointed a gloved finger and said: 'Look! Fieldfares!' I looked, too. Here was someone, elegantly poised and feminine as a suffragette, that looked out of a grimy bus window on a grey winter's day onto a brown ploughed field, saw wild winged things and knew they were fieldfares. She was excited about them – and so was I.

It was such an ordinary moment, but I wanted to be something like her, this elegant stranger on the bus who knew her birds. So, I started with the fieldfares – or, more accurately, their redwing cousins – my first self-taught lesson was how to tell the two apart. And I found I knew more than perhaps I realised. These birds were already there in my subconscious litany of things that anticipated winter; I discovered their calls and colours were old familiars, had been there all along – just un-named, un-noted. Around every Bonfire Night I could ever remember, whether on dark, moonless nights of soft rain, disappointing fog or bright, cold evenings of scarves and condensing breath, and whether we were in a town or in some other dark village, I'd always hear the first redwings come in. Such a small sound – a needling *'tseeip'* falling through the night like a tumbling pin of hoarfrost and barely audible;

the subtlest of whistles through your teeth or someone else's nose. Yet far-carrying. I did not know what they were, but I knew them. They come, perhaps, from Belarus in the night, or Poland, Greenland, the Baltics, Ukraine, Iceland, Scandinavia – and can be heard overflying towns, cities, farms, great expanses of moor or wetland. They can be heard wherever you are – bringing the colour of hawthorn berries tucked under their wing, like a portfolio of what they've come for.

Along with redwings, fieldfares are the colour of northern winters, of blank-paper skies, blue-black shale on mountainsides and weak winter sun upon snow. And a fieldfare in snow is a beautiful thing; in its element, it seems, the snow framing its aurora borealis colours like nothing else. The black T-bar tail and chestnut wings brighten against the white; its lemony, snowfield breast, speckled with hearts-and-darts, reveals a depth of the 'colour' white previously unimagined; and its slate-blue head and rump are the colour of a volcanic ash cloud. Nomadic, wayfaring (fieldfaring) and omnivorous, they travel systematically through hedgerows to make Viking raids on berries and fruit. When that sustenance has gone, they make military-style gleanings of fields for invertebrates and earthworms in loose lines, all moving the same way in the less-orderly company of redwings, dark-billed Scandinavian blackbirds and starlings that swell our year-round populations, all of them trying to manage February's cruel hungry gap.

Until then, I hadn't anticipated the coming of these Nordic raiders. Spring was the season for arrivals, when I waited for swallows, swifts, sand and house martins, annually testing and quickening my ability to tell the difference. But now when that note comes needling through the sky, it brings instant gladness and recognition. Metaphorical swallows of a darker, more difficult season, they herald a new kind of comfort: of

The Thing with Feathers: Fieldfaring

fires, frost and fairy lights, cosy jumpers and the freshening exhilaration of being out in the dark – as well as the miserable ache of a chilly house, of washing that won't dry, and mud. Fieldfares enliven playing fields and rugby pitches on the edge of towns, as well as the cold, screw-turn clods of ploughed fields, with their upright bounce and low runnings. If you were to reproduce the pattern of their movement, it would look like the charted beat of a heart over the flatline of winter soil: the earth is still alive. Fieldfare flow like pulled ribbons through hedges with quicksilver flashes, the long, cream archery of an eyebrow and a glimpse of the colour of light on snow, shaking, pulling and twisting the berried jewels from the low trees.

Unlike the redwing's thin, reedy note, the fieldfare's dialect is hard, full of harsh rattles and stony '*tchack, tchack*' contact calls. But they also use a softer, clucking '*chack, chack, chack*'. On certain gloomier days towards the end of summer, usually when washing last-minute school uniform and PE kits that haven't been replaced, the dial on the washing machine pre-empts their arrival. Three clicks for a 40-degree wash imitates it every time I switch it on: '*chack-chack-chack*'. Suddenly, in a sensory trick that triggers synapses, I sense autumn before it comes, with a wave of nostalgic anticipation and back-to-schoolness, mingled with the scent of washing powder. Then, that most domestic of sounds echoes the birds all winter – and makes me either glad to be indoors or long to be out there with them, no matter how hard the rain or cold the fields.

The direction they come from always catches me out, despite the imitative warning notes from the washing machine. I am still skewed by summer, looking south and west, to where the swallows and martins came from and are returning to. But here come the winter thrushes anyway, from behind, perhaps having taken the long route round on the return-swirl of a depression coming up from the Azores, proving their migration by the direction they've come from: of *course*, it is north and east. But for a moment, it is disorientating, the pivot point in the year, as if someone swirling water in a pan has stopped to swirl it the other way. Sometimes, when there is a big moon low in the sky, I have looked through my binoculars and seen redwings crossing the Sea of Tranquillity, the embers of a small fire glowing under their wings for warmth.

Four years after the bypass, in my birthday month of April 2000, I am back on Greenham Common, fingers hooked onto the cold mail of the chain-link fence, ready to push it down and knowing full well what this feels like. The winter thrushes have gone north a month before and I am listening for spring birds: the sweet, wild, descending notes of willow warbler above the church-bell peals of resident chaffinches or the first chatter of swallows or martins. I am not alone. We are here to cut the wire.

It is three years since West Berkshire Council bought the Common back for £1. Five years of concrete and tarmac crushing and bioremediation to make the Common safe enough for public access again. And here we are. The last of the Peace Women have left and the fences are coming down. Children from my old primary school have been invited to cut the wire and, holding hands with Martin, my husband of just seven months, we step out onto an ancient common regained. We walk onto 1,200 acres of 'new' nature reserve, against the odds, with hundreds of others, setting foot where no civilian has been legally allowed for 50 years and where, just over a decade ago, you could have been shot for doing so.

The Thing with Feathers: Fieldfaring

Heather is already shooting up from the footprint of the old runway and yellow-gold, coconut-scented gorse is in flower. During an initial guided walk round with the wardens, a Dartford warbler appeared on cue, as if it had a sense of occasion; perky and bright on top of the gorse, tail flicking, ruby-red eye glowing like a nightjar's in torchlight. After a while, we broke ranks and ran, intoxicated with the liberty and shock of it all. I towed Martin down the gullies and into the woods, kissing him in delight, rediscovering old paths and bridleways I'd thought lost to childhood. I showed him the hoofprints we'd made in protest, riding on the not-yet-set new concrete path. And then we approached the bunkers. With a handful of others, we squeezed through a hole in the fence peeled back like a curtain (or an invitation) and walked right up to the great, blind cave mouths of the hangars. No one stopped us. We climbed over the roofs of their grassy, unkempt, flop-haired tops, gripped the railings above thick metres of nuclear-impact-withstanding concrete and shock-absorbing sand, leaned over and looked down. It felt victorious, vertiginous. Like we'd won a war anyway. Though all the time, I couldn't forget that the rubble lifted from the longest runway in Europe was lying under the most controversial bypass in Europe, not three miles away. We came down, explored the thickness of the bunker walls – and tentatively went inside.

The cool, dark density of the bunker was at first earthy and reverential. Not unlike entering the Neolithic tomb at Wayland's Smithy on the nearby Ridgeway or Kennet long barrow near the flat-topped, man-made pyramid of Silbury Hill, not far away. The counter-intuitive narrow smallness of the inside *was* pyramid-like, but it mingled with a creeping, unnerving hostility: the frightening, cold dripping of underpasses, of brutalist car parks and Second World War pillboxes. There were empty gun cabinets, their doors swinging open on whining hinges; impenetrable, vault-like locked doors halfway down sloping, narrow tunnels; and the astonishing evidence of the incursions of Peace Women, in the form of graffiti on the walls. Some showed through where they had been hastily painted over; other words were fresher, perhaps added after the United States Army had gone.

The atmosphere dampened our joy for a moment. Pulled us back into that adolescent underworld our generation shared, fringed by fear of nuclear annihilation. Martin and I had been closer than most, living here. He had grown up in Penwood, a few miles from Greenham, and the village of Tadley, near the Atomic Weapons Establishment at Aldermaston, where the UK's nuclear armoury was designed and manufactured. We shared resurfacing memories of watching the animated Raymond Briggs film *When the Wind Blows* in 1986 and spoke, for the first time, about its impact. The contrast of a cosily drawn retired couple surviving the initial effects of 'Mutually Assured Destruction' was devastatingly hard-hitting.

Such powerful mixed emotions assailed me that day. We stepped back out into dazzling light, sparkling and rebounding off pools on the Common, suddenly planning a future beyond having got married; talking, for the first time and shyly, about children. Then, near the 'fire plane', rusting above its pool of water, in the place where it all could have gone off – an early nightingale sang. It was the first I'd heard since the bypass had been built. The song was uncertain, hesitant and rushed – breathless almost – as if the bird had just alighted. And of course, in daylight, the singer did not have the arena to himself. Nevertheless, it was a nightingale.

The Thing with Feathers: Fieldfaring

I broke my heart for those sudden, loud, long drawn-out notes, the quickening *'jug, jug, jug'* – and I sobbed.

In those moments, I saw what we had. I knew how things were lost and, at that moment, I felt enormous surprise and gratitude that we were there at all: that we hadn't been nuked, incinerated en masse or killed slowly by radiation poison. The heath, un-blasted. Of course, newer threats are taking us even closer to the midnight of the Doomsday Clock, in the form of climate change and global warming, but at that moment, I felt an overwhelming wave of relief, that it was possible for our wildest dreams to come true: Greenham Common could so easily have become the next biggest airport in the South, a new town or industrial park.

The collaborative determination to turn swords into ploughshares, a Cold War nuclear airbase into a nature reserve, is a testament to the will of the local people and all involved. Re-creating a living, linked, wild landscape that includes and welcomes people – and their dogs, bikes, horses, picnics, kites and buggies – has not been easy. Our relationship with Greenham Common has been fiercely won. To have it returned was a consolation of sorts, a healing, a coming together – a cause for cautious glee and celebration. The war is not over at all, but battles can be won. This was a conscious re-wilding: a re-gifting, a salve to an open wound.

Here's to the freedom of the heath.

I redoubled my efforts then, to relearn and remember forgotten species, and determined to learn new ones. At the very least, if there might be children, I owed it to them – but I also owed it to myself, to my childhood, my country, family, neighbours, acquaintances, people I would never meet. I owed it to those species that were lost and would be lost – as well as to those that would be just fine. I had a responsibility. But I also had a fierce desire. I puzzled over the importance of names at first. I could still pull the flowering grasses between the crook of my first finger and thumb to form mini trees and know exactly how each grass would feel, would pull, would scatter or would slice into the joint of a finger – an intimacy I'd not lost since childhood. In its simplest and most human form, being able to name something makes it personal and meaningful, known and powerful. Naming something is key, *a* key, *the* key: identification guides start with a biological key that helps determine known species – and we go from there. The door opens to other things, gives access. It is the start of a long look down an open-roofed and seemingly infinite corridor opening up onto room after room, or rather, farm fields, with gate after gate; a line of silk you might walk in a web that connects the named thing to other species and habitats as well as moments in our history, heritage and culture – whether on a personal or national level. Knowledge increases, curiosity is piqued, care and attachment happen, and levels of protection are provoked and invoked. Once you know the name of a thing, you are on your way to a deeper relationship with it: its habitat, what it needs, why it's there and how it connects with other things you will name. It becomes 'yours' in a way that is at once deeply personal and utterly 'owned', but also shared with and given back to the world. I began to realise I needed to learn to identify and name things better.

One day, up on the downland with my dog, the footpath took me over the vast brown plough of an arable field. It was bleak, exposed and empty, save for some rooks and jackdaws scattered on the surface like raisins on a fruit cake. Then, from under our feet, a flock of perhaps two hundred birds rose from nowhere and zigzagged upwards in a great flowing movement before sweeping past in tight formation on fast, swept-back wings. There were melancholic, plaintive cries, like the wind through a gate, before they seeded themselves back onto the earth like rain and disappeared. I did not know what they were. I stood, rooted, feeling utterly euphoric, but also as if I'd missed grasping a golden ticket fluttering past my outstretched hand.

I felt I had been dumb before. A passive observer, guilty of a Keatsian indolence. I know now that they were golden plover. A 'dread' of golden plover. The world and all its possibilities came alive and grew like a complex and never-ending lyrical poem of moth names and grasses, of scat, jizz, flint, feather, blossom and bone, and with each breath or page turned, I also found these bright new things under threat. I had to accept that this relationship would bring delight and pain in equal measure. I was always going to be hurt by this, as any consequence of opening up to love. Nevertheless, I had to own this feeling with all its responsibility, and deeper knowledge and engagement was a way I *could* own this. I was all in.

A couple of months after the revelation of golden plover, I find myself lighting fires, late March at noon, and writing an ode to a bus station. A persistent cough has become a chest infection and against my nature, I stay in, watching snow swirl on the winter wheat field behind the house and noticing that the winter thrushes have gone. I'd heard the iconic Winchester Bus Station had been included as part of the City's Silver Hill redevelopment. Its little Stagecoach enclave – tucked behind the peeling, confetti'd glory of its white plaster, prewar archway – has been a touchstone of memories, a gateway to the world and the road less travelled.

My earliest memories are of that bus station, of walking beside my brother's pram along the bright chalk River Itchen, writhing with green lengths of water crowfoot and grey wagtails, on the way to meet my grandparents off the bus. Later, it became a halfway meeting place between our homes and where Grandad would tell stories of his spell on a farm in Canada in the 1930s. At the end of the 1980s, I wrote him letters from window seats of Greyhound buses, sketching antelope on the plains of Saskatchewan for him – and telling him my own Canadian tales of cowgirl farm life, of prairies and mountains, of brandings, roundups, rodeos and coyotes; of monarch butterfly migrations, mountain lions, bears and wild horses; and of sitting on the roof of a clapboard house watching (and hearing) the northern lights crackle and wave above my head.

From the steps of the guild hall opposite, the bus station entrance framed the backdrop for my graduation and my brother's leaving party for Australia. But I also think about two girls on a bus bound for Winchester, somewhere out near rural Andover, who never knew or spoke to one another, but who made a brief connection over fieldfare in a brown and tawny winter stubble field. Part of me will always be running for the bus and a world where time and responsibility are suspended and all that matters is the travel away from home and back again – like the crochet loops of a spotted flycatcher – and the view from a window seat. But I knew what I wanted to do with my 'one wild and precious life': the challenge with which Mary Oliver ends her iconic poem, 'The Summer Day'. I intended to save, champion and celebrate the wildness in any way an ordinary girl could.

CHAPTER SIX

A Library of Landscape

What *do* you do when you no longer recognise the place you grew up in? When it has been flayed and torn off the surface of the earth; burnt, excavated, heaped up and built on with structures you struggle to make sense of? This feeling of grief and disorientation was new, distressing and seemed to permeate everything. Place was everything to me. I had been uprooted before but now, it seemed, the very place I stood upon was torn up by the roots.

By the time we lived at Greenham, I'd been to five different primary schools. By the age of 15, we'd lived in six different houses. By 34, I'd lived in a further eight houses (at one point, moving five times in eight years). Although the moves were mostly local, a few miles either side of the border between North Hampshire and West Berkshire, I didn't take this well. 'Local' meant walking distance and the intimacy that brings in knowing a place. In each place, I built an intricate, seasonal cartography of connection; of memory, belonging and meaning, signposted with flora and fauna, and carried it around inside me like a compass of home.

Moving was a necessity of Dad's job and a hardworking determination to improve our lot – which he undoubtedly did. But for me, place and belonging overrode everything; all

I wanted was to stay. I wanted roots, generations deep, and a sense of security. I wanted to know and be known. But I also missed the camouflage and solitude that come with knowing a place. I needed to know the edgeland around the houses and the land beyond better than anyone else: to go to earth and be able to lose myself to people – to remain an outlier to them. The more we moved, the fiercer this feeling became.

The moving never got easier. I felt a sense of derealisation from the upheaval – and little excitement in the prospect of exploring new places. I am certain a barely contained arachnophobia and a wariness of unknown, vengeful ghosts stem from an anxiety of unfamiliar houses and places that might not want me. I hate being the new girl, the tourist. I do not want to be caught ignorant or trespassing; to say or do anything insensitive, to be too loud or noticeable. My first defence is to read a place. I arm myself with books, poetry and pamphlets about where I am. I want to know all about a place's human and natural history, revising in case I am tested. With words, I begin to feel articulated, orientated and anchored. Rooted. And with that comes the tentative enjoyment of exploration – then I find that I fall in love with places (often deeply) – quick as that. I build a library of landscape.

Nature, literature and place twine like wild clematis and ivy through a quickthorn hedge for me. Inseparable entities: human and natural histories wreathing together like the blue-grey smoke of old man's beard through the thorny, damson-coloured density of blackthorn. The small brown *Observer's Book of Birds* that my grandad gave me, and those beautifully illustrated Ladybird books of British Wildlife and of the Seasons, were my first maps of a more detailed landscape. The animals, birds, trees and flora didn't own the place, but they belonged to it utterly. They defined it. When

I took them and their images outside, the real world popped more vividly in fully sensory detail because what I saw had been corroborated by an 'other' – an author that could validate, illustrate and illuminate my experience. Better still, fiction (or non-fiction) made a deeper, intellectual, literate and romantic connection with place. In his book, *Last Child in the Woods*, Richard Louv maintains that 'reading stimulates the ecology of the imagination.' I could (and can) attest to that: it lent me a heightened and unlimiting awareness of my surroundings, an insatiable curiosity for the natural world, and a sensitivity for sensory detail. It also connected me with who had gone before, real or imagined. From then on, every landscape became a narrative one, full of stories that could also be *my* stories. I cross-referenced, read between the (contour) lines and meandered down paths in sentences and paragraphs so that place became as much a figurative and imaginative landscape as a real one: an intoxicating, life-affirming mix of mud, bramble arches, grass, glades, woods, lyrics, paper, paint and poetry.

From the water meadows of Winchester and St Cross near where I was born, we moved, via a brief sojourn in Tilehurst woods near Reading, to the water meadows and primary colours of Pangbourne. I'd take the long route home from school through 'The Moors', the wet oozing through my brown school shoes and soaking up the white litmus of my socks, turning them brown too. I'd pause to watch water voles ferry buoyantly across the clarity of the chalk stream from the low, narrow bridge. I drew and I read. A bookish child, when I'd exhausted the library, I read books from an earlier generation, gleaned from the village's tiny Oxfam bookshop: Monica Edwards' *Punchbowl Farm* stories, Denys Watkins-Pitchford (BB)'s *Wild Lone*, *The Little Grey Men* and *Down the Bright Stream*. I read Kenneth Grahame's *The Wind in the Willows* and revelled in images I recognised. The book shimmered with the light off water, reflected by leaves. The words smelt of spring-warmed grass, cowpats, wild garlic and floodwater silt. They rang with the resonance of sunburn and nettle stings on wrists and thighs. Images spooled from the scribbly flight of slender-nibbed, peacock-coloured, banded and beautiful demoiselles, the Parker pens of the dragonfly world, writing in invisible ink above the chalk stream, glittering some days, dulled to a tin-gleam on others.

Comforting, charming and oft returned to, there is nevertheless a disquieting undertow in Grahame's book. A 'divine discontent and longing' to be out, alone, that haunts me still and periodically threatens to derail everything, including family, work and relationship commitments. Nostalgia, yearning and a melancholy air pervade *The Willows* like river mist, particularly in the otherworldly chapter 'The Piper at the Gates of Dawn', set on Midsummer Night's Eve.

The hunt for a lost otter cub recalls the naturalist's search for an elusive creature: 'they hunted everywhere for the cub without finding the slightest trace…in this silent, silver kingdom, [they] patiently explored the hedges, the hollow trees, the runnels and their little culverts, the ditches and dry water-ways.' The enigmatic character of Otter is gloriously ottery. He has a minor (though important) role in the book and is blessed with recognisable lutrine traits: he is fierce, defensive, self-possessed, private, cheerful and independent – and disappears at will, often mid-sentence.

I longed to see otters. But by the 1970s, we had lost 96 percent of our national otter population, largely due to the flush of a chemical romance with new pesticides developed during the Second World War. Chemicals that spotted our flood-watered legs in mysterious rashes and did for the fly life, fish and otters, poisoning the rivers. Yet when I return to The Moors now, I have a better chance than I ever had of seeing an otter – and almost no chance of a water vole. Like the return of the red kite, the return of the otter is proof that conservation works, but the dramatic decline of the water vole (nudging the same 96 percent in the time it took the otters to come back), for different conservation reasons, is telling. There has been a kind of vole reversal, if you like.

In his later life, Kenneth Grahame was one of Pangbourne's famous literary residents, dying at his home less than a mile from mine, 46 years before I lived in the village. Not knowing this, I assumed into my teens that I'd met him on several occasions. Dressed in a smart cravat and with a neat moustache, 'Mr Gentleman' would cross the bridge, watch the 'water rats' and mention perhaps what birds he'd heard and ask what I had seen. He once told me he should never have called them water 'rats', that it gave the wrong impression: that, of course, they were water *voles*.

As a boy, Grahame was dramatically uprooted from his idyllic Scottish mountain home when his mother died. Through depression, grief and alcoholism, his father became incapable of looking after his young family and they were sent away to live with Grahame's maternal grandmother and uncle in Cookham, Berkshire. Grahame grew up with a sense of childhood loss, loneliness and betrayal that never left him. He sought solace in the countryside, in his 'wildwood' and the Thames tributaries of the Lulle and White brooks, and he grew up something of a pagan. Dreams of studying literature at Oxford were shattered by his pragmatic, arts-averse uncle and he worked in the Bank of England. He managed to write and publish to acclaim anyway (before *The Willows* was written) and was considered a young contemporary of Oscar Wilde. But the idea of 'home', the loss and remembrance of it haunted him. In *The Willows* this is particularly evident in the occupation (and subsequent reclamation) of Toad Hall and so poignantly in the Mole's sudden, paralysing bout of homesickness. Brought on by an unbidden sense and smell, the Mole's home is deliberately 'sending out its scouts and its messengers to capture him and bring him in' – and Mole is completely undone.

———

In my pastoral, willowy idyll, all was not well – though I barely knew it. The rivers and fields were poisoned with the poorly understood agricultural chemicals in huge, unnecessary doses, hedges were being torn up and meadows ploughed under in the name of Agribusiness. Men were burning the statuesque elms, their cause of death written in the pale papyrus of their curved, cast bark in beetle hieroglyphics: the tunnelled, linocut galleries spraying out like the imprint of a fossilised trilobite. I picked them up and read them like books, tracing the tunnels with a finger before picking up another and stacking them in sympathetic piles. The first wild library of my imagination. I made the connection way back then between writing and nature, and I've never let it go.

Since then, I have become, and remain, defined by trees. Wherever I have lived, it is the trees that I remember and look for first in a new place, to orientate myself.

Beech, boc, book; buch, bokiz, boka: I *love* that the etymological *root* of a tree – specifically a beech tree – is the same for a written document, or book. A tree was a book for the first writers who couldn't afford animal-skin parchment, who wrote in runes on beechwood tablets or made tender, experimental and important marks of self-expression or communication on the smooth, human-like trunks of beech trees.

There is a commonality in the world's forests in the words we give to our trees, and in how we use trees to connect and disseminate language, thoughts and ideas. Old English derives from Germanic North European and its Indo-European languages. Sanskrit derivations give us names for ash (*bhasma*) and birch (from *bhurja*, for tree). The birch trees' ribbon-curled bark comes away in silver scrolls which can be smoothed into satiny, paper-white blanks that we can write on, with the purple-stained ink of their twiggy, besom-broom branches.

Trees have their own language. But they can't be separated from ours.

I liked the poetry of that.

Knock on wood, tap on a book, and you can hear the satisfying comfort of reassurance. The authenticity of the timbre of timber: of plane, light and growth. In death, a tree comes alive again (the wood that warms you twice, or many times that). Trees give life, even in death, to much wildlife, to stories or as fuel or building materials: a whole world of imagination, practicality, engineering and culture. Our history is written down in rings of grain.

As a young child, I loved The Moors on the edge of Pangbourne. I loved the name, loved saying: 'I am going to The Moors' – though I was aware they were distinctly un-moorish. Catherine Earnshaw would have scorned the one I knew as such, though there were lapwings. It would be a few years before I'd read Emily Brontë's *Wuthering Heights*, but I already had a working knowledge of it through TV, film and Kate Bush. I knew the creak of a heron flying over. I knew how to stalk like one with my shadow behind me, to marvel at the tiny dents a pond skater made on the surface of the cattle drink, spreading its fly-weight of nothing to stand on water. Lying on my front, I watched shiny carousels of whirligig beetles bump round the kidney-shaped pools left by cow's hooves. Like many children of my generation, we were always let loose, outdoors. Mum would walk us for miles without agenda, restraint or pressure of time. And walking beside the Pang with my Romany grandad (naturalist, storyteller, something of a 'scholar gypsy') felt much like being Mole trotting beside the river. 'Mole was bewitched entranced, fascinated. By the side of the river he trotted as one trots, when very small, by the side of a man who holds one spellbound by exciting stories…a babbling procession of the best stories in the world, sent from the heart of the earth to be told at last to the insatiable sea.' When I wasn't out on The Moors, I spent weekends and summer holidays at a tiny riding school of just seven ponies, grooming, leading and mucking out for rides.

In Pangbourne, the flow out to a faraway sea could be witnessed, as the little chalk streams of the Pang and Sulham brooks fed into the wide Thames with its big boats and little beaches in the water meadows. The annual flooding was awe-inspiring. To see the river burst its banks and run faster than a man could across the pasture was both terrifying and exciting. We were allowed to wade in – but never beyond the tops of the benches, if they were still visible. The brown swirl and undercurrents of the river under the span of the white iron Toll Bridge were lethal and drowned the brother of a girl in my class at school. I felt its greedy pull once, fishing for tiddlers with Nan in high summer. I slipped from a little concrete jetty into it, knocking Nan over so that we both went in the bright water, our legs instantly and powerfully tugged from under us, into the current, as if we were flying like flags in high winds. An indomitable spirit, Nan wasn't going to let me go. She grabbed a tuft of grass and my skirt and I grabbed hers. I remember the patterns of blue paisley cloth darkening to indigo in the water, our fists clutching cotton swags and stretched elastic for dear life. Had we been boys, we would have been swept away.

When we moved again, I packed up my achingly realistic toy Britains Farm, with its little plastic post-and-rail fences, hay rake and Land Rover, and my books, and tried to get my bearings, reading alone in the playground at school as camouflage. I comforted myself with pony books, with K.M. Peyton's *Flambards* and Enid Bagnold's *National Velvet*. A self-reliant 11-year-old, I read my way through displacement, homesickness and belonging in *Anne of Green Gables*, *Little House on the Prairie*, *The Secret Garden*, *The Diddakoi* and *Heidi*. I immersed myself in the intense, sensory, ankle-height world of *Tarka the Otter* (and imagined otters in benign chalk streams), along with the unsentimental countryside novels of Joyce Stranger and the romantic countryside of Mary Webb. I found a profound, ancient and spiritual connection with place in *The Little White Horse*, *Moondial*, *The Dark Is Rising* series and in Patricia Leitch and Alan Garner's books – being completely undone by *The Owl Service* – and I avidly read the nature column *From the Hedgerows* in our local paper, the *Newbury Weekly News*. I carefully learnt the correct pronunciation of local place names ('Woody' for Woodhay, the 'Ilzees' for Ilsleys) before I spoke them, as if, by proxy, I could cover myself in their soil; gain new purpose and footings from the new landscape, its intimacies and peculiarities. We'd moved a distance of 15 miles.

But nothing maps the narrative-landscape of my life quite like Richard Adams' *Watership Down*. It is a book so local to me – I have lived in every location it maps. I grew up where it starts and made my own pilgrimage to the 'high, lonely hills' where it ends; myself a refugee from human development and a changed landscape that had buckled and bucked and thrown me off in its death throes. The story is told from a rabbit's-eye view and is as intense, immersive and unsentimental as *Tarka*. It is a dark, brutal but eventually uplifting tale and I can still recall (as many of my generation can) the haunting film of 1978. It pulled no punches. Though set in a particular natural environment, the rabbits have their own language, culture, religion and mythology. The novel is sown through with the intimate and precisely local details of wildflowers, chalk streams and birdsong that I still sing along with now; the *'little bit of bread and no cheese'* of yellowhammer song and the strident *'cherry dew, cherry dew; knee deep, knee deep'* of song thrush.

I worked at a stables in the grounds of the estate where Richard Adams went to school. Next to my old secondary school in Wash Common, a small housing estate is entered via Warren Road – a nod to the very real housing development that precipitated the fictional rabbits' flight and ignited the author's imagination. The appearance of a developer's sign pre-empts the 'seer' rabbit Fiver's apocalyptic vision which instigates the start of an epic journey. When the rabbits flee, they cross the River Enborne (its source a few hundred metres from the house I live in now) and the heathland of Newtown Common, *'where the very plants were strange to them'*, is where I got married. We lived, then, in the wing of a manor house on the heath, where my new husband worked as a gardener and a groom. Putting the ducks away at night was an adventure: via a rowing boat, manned standing up, across to a tiny island in the middle of a lake. From June to August, we were lulled to sleep by nightjars. From here, the rabbits cross the River Test, which seeped up through the cob-and-chalk walls of the first house I shared with my (then) future husband at Whitchurch, a cottage built in 1777 between houses and with no front of its own, in the little town where Richard Adams lived until he died. For a while, we lived in a burrow-like cottage ourselves at Siddown Warren, opposite Watership, on those *'high clear downs'*. I have ridden, cycled, walked and even slept out on those hills. I know the title of Chapter 5: 'The North-East Corner of the Beech Hanger on Watership Down', intimately.

And the connections and re-connections keep coming. The book is a touchstone. Forty-four years after it was published, Mum took my two daughters and I to watch *Watership Down*, the play, adapted by Rona Munro at our wonderful, tiny, warren-like Watermill Theatre (one of the smallest in Britain). The Watermill was the scene of one of the last, intense bypass battles, where the Goliath work was almost stopped by the tiny David Desmoulin's whorl snail and where a nightingale sang fitfully from the reeds.

The story was beautifully and cleverly told. Richard Adams was able to watch it before he died later that same year on Christmas Eve, aged 96 – and reputedly loved it, as I did. The theatre backdrop could have been a warren entrance, a shining stream or a chalk track, nobbled with flints and more luminous than the sky, leading to home and the hills we run to. It was infused with a warm, warreny push-and-pull of cosiness and claustrophobia, illustrated with utilitarian wartime austerity: a stoat puppet was frighteningly evoked with a gas mask and concertinaed tube. Sometimes it takes a new generation to discover an old classic. On rereading it as an adult, the battles, protest and odyssey take on a fresh and poignant meaning and relevance. The rabbits are threatened, oppressed, hunted refugees who ultimately make it to their place of greater safety and a new life. The book is about exile, community and home. The opening scene is played out below the centrepiece of a developer's sign, proclaiming the coming of new houses, detailed at the beginning of the opening chapter. The irony was almost too much. In reality, this ex-warren, this former playground of mine was to be further devastated by an even bigger, new development of a proposed 2,000 houses. Art imitating life imitating art? I wondered how many in the audience realised? Waiting for curtain up, I wanted to stand up and shout about it. But I also wondered if Adams knew what binding, freeing magic he made, writing a story with such a strong sense of ecology and place? Its journey, earthbound premonitions and its sense of things beyond sustained me. As with the map in my Hardy novels, I trace the real landscape through the imagined; words and sentences were my contour lines, keys and grid references, images my triangulation points, and chapters my northings, eastings – my legend.

―――

I like to read a book 'in season' or 'in place': rereading *Wuthering Heights* in a storm on the Yorkshire Moors on holiday, *The Return of the Native* on a glow-worm-peppered heath (where I can still, just about, find them) and *Tess of the D'Urbervilles* with my back to the gibbet on the high chalk hill above our home, with distant views towards Winchester Gaol and Salisbury Plain on a good day. Phrases come back to me in certain months or weathers, or emotional circumstances, as if they were written on the landscape for me to read, in sharp relief. Thunderlight on the plough, dusty nettles in the farmyard, winter rooks or the silhouette of the barn and the down above it, all trigger lines from sometime-native poet and writer Edward Thomas. I search habitually for the first sweet white violets in early spring (which I also mistake, as he did, for a 'chip of flint, and mite/Of chalk', in his poem 'But These Things Also'). The words come with the seasons as surely as the mistle thrush's melancholy carolling after December 21, or the first spilling over of a full phrase of chaffinch song as the days continue to lengthen. I feel them profoundly. With them, there is a deep connection to the rural past and with those that have lived and worked here before, as I think Thomas felt or searched for. I recognise in his lines their ghosts; the tenacity of their continuing, contemporary presence and influence: in the swirl of dust-devils in the yard, wild mignonette growing out of the corner of the farm estate office building or the way carted straw still banks the lanes and hangs, festooned like golden bunting from low trees at harvest time. Thomas knew this. His writing is full of a sense of longing and yearning for something that exists just beneath the surface. A recent past perhaps, or a land he never knew, a certain kind of grief or nostalgia for something lost or just out of reach: land, history, home, the connection between people. Thomas connected strongly with Wales: his father was a Welsh speaker and there was family across South Wales where Thomas spent many of his childhood holidays. The Welsh have a word for this feeling: *hiraeth*. A deep longing for a homeland, more acute and nuanced than homesickness – and not necessarily for a home you are from or, possibly, that even exists. Its meaning is distinct from *cynefin*, another Welsh word without English equivalent, which is more a physical and emotional attachment to a place or particular habitat.

Hiraeth is what American writer and artist Pamela Petro calls an ecological keyword, 'an intellectual and moral home' and (importantly for me) 'a protest'. Writing on the subject in 'Dreaming in Welsh', in *The Paris Review* (2012), Petro describes a feeling familiar to her 'because home isn't the place it should have been. It's an unattainable longing for a place, a person, a figure, even a national history that may never have actually existed. To feel hiraeth is to feel a deep incompleteness and recognize it as familiar.'

As the ecological crisis deepens, new words enter our language. The word 'solastalgia', coined by environmental philosopher Glenn Albrecht in 2005, describes the disorientating grief and emotional, even existential, distress caused by negatively perceived environmental change or destruction – and it can be applied to many situations: the mining of ancestral lands, house building on a childhood field, the felling of a much-loved wood, even the loss of 'proper winters'. Albrecht describes the effect as a homesickness, for the place that once was and that you haven't left. Ginny Batson, writer, eco-philosopher and environmental ethicist, has created many neologisms: new words where we need them, to explain, understand and ultimately improve our connection with nature and the health and well-being of the planet. Among them is 'fluminism': an interconnected narrative of a dynamic universe; a complex, living flow between humans, the planets and nature that we're only beginning to comprehend. We have little or no language for these unprecedented times of ecological and climate breakdown, and admitting and owning a love, as well as a grief, for the loss of nature is key. And of course, it begins with language and naming. When I am feeling shut in, cut off from 'out there', upset or feverish, lines from *Wuthering Heights* haunt me as manifestly as Cathy's ghost: I want to be away from where I am wanted or expected to be, throwing open windows and running to the hills: 'heaven did not seem to be my home; and I broke my heart with weeping to come back to earth; and the angels were so angry that they flung me out into the middle of the heath... where I woke sobbing for joy.' Like Shakespeare's drowning Ophelia, garlanding herself with the wildflowers she carefully names, Brontë's Catherine Linton feverishly tears her pillow, and plucks and names the feathers she finds within: 'here is a moor-cock's; and this – I should know it among a thousand – it's a lapwing's. Bonny bird; wheeling over our heads in the middle of the moor...I'm sure I should be myself were I once among the heather on those hills.' This is Catherine's *cynefin*. Her *hiraeth* (and what could be interpreted as her cause of death) Heathcliff and the moors between Wuthering Heights and Penistone Crags – not as part of her feral, brutal childhood, but as they might have existed had she chosen him over Edgar Linton.

I go out to regain a sense of myself: of space and calm – or wildness. Whatever I need, it is always there. My chalk hills form the feminine curve of the downs. I lie down, fitting my hip into its smaller curves and resting my head on the grass pillow of a meadow anthill, cushioned by wild thyme and fairy flax, and breathe in a wild garden of herbs – basil, marjoram, salad burnet, calamint. Like Seamus Heaney's *Antaeus*, 'I cannot be weaned/Off the Earth's long contour'. Like Antaeus, that is where I find and regain my strength.

Just four years before the bypass was built, Adam Thorpe's classic novel *Ulverton* was published. It is centred on a fictional amalgamation of places he knew: the old downland villages around Newbury, using a version of its first recorded name of Ulvitrone. Using themes that twine like red ribbon or wind like the smoke of wild clematis (old man's beard or 'bedwine'), 12 stories are told about a place through over 300 years of English rural history. The stories are excavated or found fragments, a narrative archaeology, not always easy to decipher – some stories are told in dialect, as a rambling sermon or pub conversation, a script, as one side of a correspondence or written half-illiterately – but the voices are utterly authentic. It is a haunting story of place, told by its inhabitants down the years.

Ulverton begins in 1650 with a shepherd encountering a villager returned from the English Civil War. The soldier finds his wife remarried and promptly vanishes. In 1989, his remains resurface, threatening to derail a lucrative housing development proposed by the second husband's descendant. The landscape speaks for itself – is given a voice. Between these bookends, voices come from gentleman farmers bent on improving land and legacy, letter-writing women from either end of the social spectrum (but each confined in their own way), ploughman, gatemakers, machine-breakers and swing rioters, photographers, archaeologists and documentary makers. A white horse is sculpted out of the chalk, a barrow excavated, young men are encouraged to war. There is a link with Egyptian archaeology, mention of a motorway protest and a housing development in a rural location that Adams' rabbit Fiver would have baulked at. It is a story that is intrinsically local; a core bored through a rural, human presence as an increment auger through a tree – that could

A Library of Landscape

equally be applied to any piece of British landscape. *Ulverton* is a kind of literary archaeology; a stratigraphy of voices that thread through each other from the deep past – yet that a gentle tug would bring right into the present, as happens in the final chapter. Thorpe describes his 'idea that my book's hero would be the place. It came up to me from the earth, it really did. The energy was coming up through my legs.'

Thorpe's anatomy of a Berkshire village is deeply affecting, relatable and recognisable. It pieces together a story from the unrecorded fragments that surface or exist in a place: a knapped flint, potsherds, a horseshoe nail, a bleached piece of red ribbon or the way hands have rounded and smoothed a gatepost. Ulverton is a fictional village, yes. But I know it. It is home.

A literature of place and home gets you in and takes you deeper – and in the absence of roots, a book is an uncanny anchor. And if reading explores the question 'Where do I belong', then writing explores the '*How* do I belong?' Adam Thorpe writes himself a cameo role in *Ulverton*, as a local author and dramatist.

———

I work now as a librarian at the small, rural comprehensive school in Hungerford, where straw blows under the sliding doors on some days and the view of the downs is best viewed from the classroom in which former student Michael Ryan shot himself, after killing 16 people and injuring many more one summer Market Day. I strive to get those children reading who would rather be out in the fields, farms or on the Common. I love that the word 'library' is rooted in the Latin *liber*. The same root word for book, tree bark and freedom – liber*tree*, if you like.

While we don't study Latin, of course, I carry the knowledge, freedom and *treeness* with me when I go to work. Rooks from the tall limes opposite herald me in and barrack me out again at dusk. Our school writing club is called The Rookery Writers.

I've always written. Move me and I'll start a relationship with a place immediately, reaching out tendrils that curl into words and wrap around syntax, with an ache for home and a need to get the dirt, moss and grass seeds of the new landscape under my fingernails and onto a white page.

I cannot experience nature without it coming out in words, similes and metaphors. I am compelled as a reader and a writer to try and reproduce, record, interpret and hold the experience in my senses, fresh as the moment it occurred. It's a kind of ownership in itself. A capture. A possessing and a possession. And it's how I orientate myself. Sometimes, this means I am lost. Not to myself, but to others. Tenant, flitter, plover, I can be found roaming the down, my head stuck down a badger sett, feet in a furrow, hair tangled in the hawthorn, following a hare. Or on the hill shaped like a loving arm, the combe, a hollow beneath a collar bone. My library, my ivory tower, is a chalk hill of stacked geography. I converse with nightjars and green plover, writing the calligraphic curve of crest and looping flight, grown cursive. When I lay my head on the hill and breathe in an old oceans' worth of loving, crushed basil and wild thyme, I glean words, not grain; I make sense.

CHAPTER SEVEN

Tenant, Tied

As tenants or workers with tied accommodation, my new husband and I moved and moved, hopscotching along the West Berkshire and North Hampshire border. We took up one short tenancy in a converted stable on The Ridgeway after a horse my husband was working with kicked his hand and broke it. Unable to work, he lost his job. Not long after, I fell pregnant.

Seeking some security, my husband took employment as stud hand at Highclere Stud, a position that brought with it a house. In sight of Watership Down, Highclere Castle and its Estate was the home of Lord Porchester, the 7th Earl of Carnarvon and the Queen's friend and racing manager.

Our cottage was secreted behind the high dome of Beacon Hill, down a narrow, mile-long drive. I lost my heart to it. One of a pair of traditional farm workers' cottages, it had a shared attic and a roof that sloped to the ground at the back like a sou'wester hat. This protected it from the elements, created shade for a cool pantry and a draw (through the position of the door) for the substantial fireplace. It was built, like the badger sett above it, into steeply wooded Siddown Warren Hill – historically, an enclosed rabbit warren or 'bury' in this high clearing, once managed by a manorial rabbit keeper or 'warrener'. We were never the only residents. Tiles overhung flint-and-chalk cob walls, creating all sorts of crevices for insects and bats. It had a huge wood-burning stove, and needed it; for a while you could poke a finger through the rotten window frames and the old latched door, painted in layered decades and shades of Estate blue paint, had a two-inch gap along the floor that rain and hail blew under. In unexpected snows, a drift would form in place of a draught excluder. The outside was always expecting to come in. D.H. Lawrence's Mellors and Lady Chatterley would have recognised it – except perhaps for the flint.

The cottages were sheltered by a beech wood. The towering, smooth, grey torsos of the trunks with their sky-thrown limbs dominated the approach. They were strangely human, at times stately; not unlike the Bath stone of Highclere Castle when wet. They are almost all our son recalls of his first home. It was so dark and quiet at night that it was easy to miss the path down from the roof-height lane above to the door below. In spring, when the ground was covered with a thick hushed mat of beech husks, you could walk right into your neighbour in the dark. But in winter your footsteps crunched over the muesli of cracked beechmast and gave warning. Like all beech woods, it rained out of synch with the rain. Overlapping lacquered leaves acted as umbrellas for a while, but continued a steady pattering long after the rain had stopped.

The valley widened out from the cottage into wildflower, herbal and 'medicine' pastures for glossy, impeccably bred thoroughbred mares and foals – everywhere I turned looked like an Alfred Munnings painting. The 7th Earl, Lord Porchester, was a first-class breeder of horses and an expert on grasses and herbal leys. The springy turf nurtured fast animals with tensile fetlocks and buoyant hooves. The calcium coming up through the grass from the chalk made for clean limbs and strong bones. There were hares in the stallion paddock, barn owls in the haylofts and, in winter, a stoat in ermine. The slopes of the hills were dotted with the long, shifting shadows of box, juniper and silvery flashes of whitebeam. The lower reaches of the valley and a white chalk track curved protectively, jealously, around the cottage and fields. The bank and ditch of an Iron Age hillfort crowned the hill and built on top of that was the grave of Egyptologist and founder of the stud, the 5th Earl of Carnarvon, our employer's grandfather and, reputedly, one of the candidates for the inspiration behind Toad of Toad Hall. (This is even remarked upon by his fictional contemporary, Countess Violet of Downton Abbey. In the historical television drama based at Highclere, and written by Julian Fellowes, the sharp-tongued matriarch remonstrates with her granddaughter Lady Edith, that she is 'a Lady, not Toad of Toad Hall'.) The Earl's grave was surrounded by iron railings and situated so as to be visible from both his bedroom at the Castle and the stud. Together with Howard Carter, the 5th Earl discovered the tomb of King Tutankhamun in 1922. In the late 1980s, recently retired butler Robert Taylor rediscovered some three hundred forgotten Egyptian antiquities during an inventory of the castle for the 6th Earl's son (soon to be 'our' Lord Porchester). The treasures were hidden in sealed cupboards in the panelling between two rooms, labelled in Carter's handwriting, apparently because the 6th Earl had no interest in them. A schoolfriend of mine, who worked on Saturdays in the kitchens, was shown them at the time of their discovery and sent on a search for more neglected antiquities languishing in forgotten drawers.

The artefacts had lain hidden for more than 60 years and now form a private exhibition in the cellars of the castle, open to the public and school visits at certain times. But perhaps someone knew, other than the butler. Perhaps the clues to the artefacts were there all along, in the names of some of the fine racehorses bred at Highclere: Hiding Place, Smuggler and Cubby Hole – and the famous Niche, who was foaled after the discovery.

———

At Highclere, it did indeed seem possible to hide from the world and belong to the just-discernible rural past. There was still an old rhythm to life, harvest suppers for the farm and, the stud farm being pasture, we were buffered from the various ill effects of pesticides and herbicides on the local wildlife to a greater extent. With just half a day off a week, a whole day once a fortnight and a bloody good reason needed to request a 'holiday', we rarely went beyond the Estate gates. We were absolutely tied to the accommodation, our employer – and the place. There was always mud on the bootscraper, the smell of bran mash or boiled barley on the stove and straw on the floor (as a visiting midwife noted wryly), and I made a nest. The Queen came to visit her horses on the day our son was born and congratulated my husband.

We felt, sometimes resentfully, part of the ghost of an old order of things. It was there when a low feudal mist cut us off from the outside world, when the Queen or wealthy Arabs visited, when we were allowed to collect firewood, cut ourselves a Christmas tree or accept a brace of pheasants or the haunch of a muntjac from the gamekeeper. It was particularly present at Christmas, when we walked with other Estate staff along the old Sunday Church Path through

Tenant, Tied

'The Wilderness' woods and up the mistletoe-baubled lime avenue, carrying our best shoes, for drinks at the Big House. These were relaxed, generous affairs around an enormous Christmas tree in the grand entrance hall saloon, with its painted leather wall coverings from 1631. Portraits of the 1st Earl, killed during the First Battle of Newbury in 1643, hung on the walls. When the 7th Earl died, the Estate staff and their families were invited up to the Castle for the wake. I bought a new outfit for the occasion, having nothing to wear, and worried about the extravagance of it. There were speeches and plenty to drink, and the occasion was a relaxed, respectful, jovial and fond celebration of life – and the castle glittered resplendently. As we moved from the saloon into the library, Martin became the butt of jokes for spilling his drink on Napoleon's desk.

The past was tangible. Between morning and evening stable routines, we picnicked in bluebell woods that rang with the calls of cuckoos, leaned upon beech trees scored with the names and initials of former grooms, gamekeepers, house servants and gardeners: the most moving of these carved before they went off to war. A favourite walk was a climb up the old wooded Warren above the house to Heaven's Gate. One of several follies on the Estate, it was the most romantic. Mid-18th century, the tall, red-brick-walled archway stood at the top of the hill like the last gable end of the ruined grand house it had never been. It framed a view of the Castle in its park and the landscape beyond. Carved into the soft red brick were more names and initials of former staff: workers, lovers, soldiers leaving their mark in the only way available to them. On windless days and nights, when the ease and movement of the trees should have been still, I fancied the creaking I could still hear came from old carts going up the track: a horse's hooves or hobnailed boots muffled by years of beechmast fallen on chalk.

I was determined to bring our son up with a love and understanding of this landscape and a deep sense of his place in it. The way of life was all-consuming. There was a gentle sense of healing and rediscovery for me, as a first-time mum; a hiding away in a pastoral idyll that I knew didn't really exist – at least, not fully – and possibly never had, except in reflective moments of gratitude, rest, peace and plenty.

It was a close, insular time. Our world revolved around the horses. During foaling, my husband took his turn sitting up with labouring mares in the dark end of winter, and I'd take him meals cupped in tinfoil, our son asleep in his pram by the stove in the feed room. In unexpected storms – once, memorably, when the heavens threw everything down at once: snow, hail, thunder and lightning – the yard staff ran from their houses to get the panicked yearlings in from the fields before they injured themselves. Standing at the back door, heavily pregnant, I watched a dance with death: cloudbursts unrolled like bolts of white tulle from a navy sky and forked lightning hit the ground in a maelstrom of snow swirling like a Turner painting. Each flash illuminated running horses and men and women stood like statues, one hand outstretched, palm up, the other holding a headcollar tucked behind their backs.

And then, in some distant reverberation of wars that took and shook the lives of the household, stud and farm staff before, the world came to us anyway. With the Iraq War seemingly imminent, my husband's Wednesday nights and occasional weekends training as an RAF Auxiliary reservist got serious. There were rumours that reservists would be deployed.

Tenant, Tied

His call-up papers were delivered just before Christmas 2002, as I was hauling prickly armfuls of holly into the house. The post lady hovered. 'I know what that is,' she said, with a haunted, accusatory expression. 'My first husband was called up for the first Gulf War. And never came back.' She walked back up the path through fog so thick she quickly disappeared; though her words hung in the air as if painted on a solid, creaking pub sign that had suddenly materialised above the back door. I hadn't heard her van arrive. I didn't hear it go. A fainting static began to roar in my ears and gave an unstable, migraine edge to my vision. A metallic, ashy taste crept into my mouth – and took me right back to the warning sirens of schooldays at Greenham.

After weeks of false starts, goodbyes, postponed leavings and returns, he was gone for six months. In splendid, shaky, apprehensive isolation with our 18-month-old son, Billy, and dog Tess, I reacted by digging in. I needed roots and soil more than ever to stop myself blowing away.

We explored the 5,000-acre Estate, testing the rugged-wheeled pram to its limits – or when that was not possible, with Billy in a carrier on my back. We walked to Heaven's Gate, to Watership Down, to the lake and mere, and up to the grotto and the follies – in various states of romantic but serious ruin. The Etruscan-styled Temple of Diana had last been lived in by the tenant shepherd – and was now given over to sheep. A weekly pilgrimage was a hard pull up to the summit of Beacon Hill that took me through the centre of its hillfort and the scooped, nettle-filled hollows where its Iron Age huts had been. At intervals, like the spoke-stops on a great cartwheel, steep tracks were worn through the thin turf to bare white chalk. Slick and like a smooth-sided luge-run in damp weather, I doubt grass had grown there since it was built around 1,000 BC. This hillfort, like its neighbours, must have sat spectacularly, like an enormous, white, gleaming crown on the summit of the domed hill, visible for miles around and glowing with an ethereal inner light.

From the ramparts facing west, leaning against the railings of the 5th Earl's grave, I could look over an ancient, farmed landscape of strip lynchets, field systems, barrows and terraces – and down onto the white chalk of the Estate-owned Ivory Farm, the stud, paddocks, woods, barns and stable yards, along with the tower and turrets of the Victorian Highclere Castle and our community of seven houses. This was the only place that revealed our little enclave: our house, in its encircling bear hug of hills, like a guarded secret.

I climbed the hill in the days after news reached me of the tragedy in Majar al-Kabir – when six Royal Military Policemen were killed. There was a total news blackout, and no way of contacting my husband. A Chinook on manoeuvres from nearby Salisbury Plain came up over the hill and we watched it disappear below us in the valley, my son excitedly jiggling about in the back carrier. This was a fairly regular occurrence, but on this occasion it reappeared suddenly in front of us, like a giant, drowsy wasp, and dipped its black nose in greeting, the long, double rotor blades beating the air out of my chest and flattening the grass all around. My son took in great gulps of astonished downdraft before bursting into tears; the dog stood on her hind legs and barked. We slipped and slithered down the hill towards home, breathless, exhilarated, frightened. I hardly knew whether to take it as a sign that everything was okay, or as a portent that it wasn't.

Tenant, Tied

Days later, I got to speak to my husband on our weekly timed phone call. Of course, he was fine – but he'd been shaken by events.

I sent books to hold him, to keep him safe, to ground him. I sent him (inadvisedly) *If I Die in a Combat Zone* by Tim O'Brien and Siegfried Sassoon's *Memoirs of an Infantry Officer* and also Michael Ondaatje's *The English Patient*. In the furnace heat, the glue in the binding melted. He copied out the passages describing desert sandstorms when the sun turned red and the air had the unreal, foxed, sepia tone of an old photograph – and sent them back to me. When he returned and I unpacked the book, grains of sand had become preserved in the reset glue of the spine.

We were able to email most days, but also sent each other military aerogramme letters known as 'blueys'. I have a suitcase full of them reunited: everyday stuff like news of our boy walking for the first time down to the stallion's box and his first shoes; the frustrations and miseries of being apart; and all our separate adventures. But, through the writing of letters, we found a renewed respect and appreciation of each other as individuals. An expression of physical and emotional longing – as well as a certain freedom – was set down in words with all the passionate intensity of long-term, long-distance separation, to be read and reread. It was a very sensual, mindful time.

Alone in the comparative wilderness with sole responsibility for a baby, who became a toddler, I sometimes wouldn't speak to or see another human being for days. I slept lightly, alert to any sound or instinct, like the mares in the fields. My senses were piqued, and I learnt to listen to them more, to tune in to the wild world around me with renewed intensity.

Out walking, right on the edge of where I was permitted to go, in the woods one day, I felt the air suddenly change around us, the light flicker. I called the dog to my side, hoisted Billy up higher on my back and waited. And then a whole forest came up and over the slope in front of me, in chaotic, lurching, rocking-horse motion. Branched, stag-headed like old oaks, a big herd of fallow deer came hurtling towards us, holding crowns of flat, thorny antlers high as raised candelabra, each wide enough to hold a piece of sky. Birnam Wood doth come to Siddown Warren. Bucks and does, their heavy bodies thick as brown tree trunks, leapt, pronked and galloped at us on strong, sapling legs. There was nowhere to go. We stood as they flowed and careered around us at speed on the narrow grassy ride, catching glimpses of a wild white eye, a sharp white tine, a heaving, dappled flank, a lolling tongue; a doe's mouth downturned and part-open, the hot shock of grass-fomented breath and the familiar tang of ammonia. They created a dizzying strobe effect as they moved in front of the sun, blocking then revealing it. I clutched the straps of my toddler's back carrier helplessly as we stood at the centre of a zoetrope of deer, cloven-hoofed pinches of flung mud raining down around us. They came in all colours: black melanistic, a white buck – a white hart! – and spotted chestnut. A big buck, head held back on a long swan neck to support his huge antlers, was almost upon us before he shied violently to my right and ran through the open pheasant pen. On reaching the end, I turned to see him half-leap, half-bolt through the wire fence at the other end as if it wasn't there, leaving a deer-shaped exit hole in the mesh. At my feet were the deep, wide-splayed prints of a very large deer taking evasive action – a pair of empty brackets around a sentence of speechlessness. I don't know how we weren't knocked down. I don't know what spooked the herd into their stampede, but, after the air quietened around us,

Tenant, Tied

I turned too and followed them home, trying not to run myself; the air rich with their running and scent and fear.

Evenings after tea, I would hoist Billy on the back carrier and climb the steep slope to the badger sett above the house. Dug into the side of the hill, just like the house, we could look down onto the sloped, mossed roof through the trees. The lane became obliterated by the tree canopy and the house became, satisfyingly, part of the hill, the badger sett simply an upper floor. I had dug in; made a nest, a sett, an earth for our tiny family. Dark, warm, small and cosy, with a bright fire and flint front opening out onto a sunlit combe, its deep shadows were as meaningful and rich as a Ruralist painting.

Watching badgers wasn't (perhaps understandably) that successful, with a very small child. But sometimes, we sat among it all and were calm; muntjac deer got used to us, walking past with furtive, apologetic, pig-like movements, and once, we watched a stoat hunting and a woodcock feeding, methodically depth-plumbing the earth with its long bill. I liked to just be there, taking note of the deep scores on a bent elder tree that the badgers used for cleaning and sharpening claws, the bridge of its underside greased and stuck with grey, black and white hair where the badgers had scratched. I would sit, roll and bump their wiry, square hairs between my finger and thumb. The sett was probably a much older ancestral seat than the Castle and their highways worn through the green unfurling of dog's mercury and later bluebells were as ancient a track as The Ridgeway.

Sometimes, with a cuckoo calling well after bedtime, we would hear badgers bumping about in the ground below us, or the sound of them bickering. Occasionally, a humbug-striped face would appear from the soot-dark holes, sniff the air and retreat. Or a silvery lozenge would materialise from an unexpected direction and pour itself like a slippery fur rug down a hole. Once, a big sow emerged into the lemon evening light just feet from us and scratched audibly. Two cubs came out with chalk smudges on their noses and climbed her back like bears, falling off, chewing at her ears. A wave of Billy's excited arms in his rustle-y outdoor all-in-one and they were gone. Often, I would stay too late and walk back down the steep, pathless slope in the near-dark, Billy asleep as a deadweight on my aching shoulders. The light thickened and played tricks with my eyes. Pale blotches of lichen on the trees seemed to come loose and float in the air; the light strip of a low, lacerated elder, a chalk scar on the ground, elderflower heads, a flint or a large, blundering white moth – all could be badgers. Once, a disembodied 'V' of stripes *was* a badger's mask, which seemed to hover in the dark before vanishing on the path down to the house. The wood was alive with badgers, just out of reach.

In 1821, William Cobbett began a tour of Southern England on horseback, writing about his series of *Rural Rides* in the *Political Register*, a journal he founded that became the main newspaper read by the educated working class. Cobbett campaigned for free speech and was sued for libel on many occasions. He was tried several times for sedition, spending time imprisoned in Newgate as well as in exile. He was a complicated radical – with a hatred of injustice, a love of tradition and a deeply conservative instinct. But in particular, he was a determined champion of rural workers, and of the rural England they had created over the centuries. He was a campaigner who openly derided 'The Establishment' (whom he referred to as 'The Thing' or 'over-fed tax eaters'). He hated the game-preserves of grand estates such as Highclere and the privations caused by enclosure and land ownership, which in effect had turned rural labourers into slaves, denying them the sustaining traditions of gleaning, commoning, gathering and working for a fair wage. He rode to see for himself the lives of the common man, woman and child, avoiding the detested turnpikes and gentrified smoothed roads. He chronicled, wrote and 'harangued' (in his own words) his way around his beloved countryside, acting as a kind of roving representative, rallier and spokesperson of a changing English countryside, often sleeping on the road even into his sixties and giving away whatever budget he had left after meals to those who needed it.

As he rode through my home counties, Cobbett witnessed the more devastating effects of the Agrarian and Industrial revolutions that I had studied in History at school. Even as he became an MP, he risked jail and deportation for his support and encouragement of the Swing Riots taking place at the time. His journals, collected into *Rural Rides*, have never been out of print.

Cobbett liked Highclere's setting very much. In his journal entry for 2 November 1821, he wrote: 'This is, according to my fancy, the prettiest park that I have ever seen. A great variety of hill and dell...I like this place better than Fonthill, Bleinham or Stowe, or any other gentleman's grounds that I have seen...the great beauty of the place is the lofty downs, as steep, in some places, as the roof of a house.'

My house.

Cobbett also notes: 'Our horses beat up a score or two of hares...from the vale in the park, along which we rode, we looked apparently almost perpendicularly up at the downs, where the trees have extended themselves...as variously formed glades. These, which are always so beautiful in forests and parks, are peculiarly beautiful in this lofty situation and with verdure so smooth as that of these chalky downs.'

Lord Carnarvon did not like Cobbett's politics and Cobbett was not impressed with Carnarvon's Castle: 'The house I did not care about, though it appears to be large enough to hold half a village.' There were things in Highclere (as well as in any large country estate in Britain) he would have loved and recognised – and been incensed by, also. I am in no doubt he would have recognised the continuation of aftershocks that the Enclosure Acts and the Agrarian Revolution caused.

Cobbett was incorruptible, yet unaware of the dangers of nostalgia and, most shockingly, bizarrely and counter-intuitively, a terrible racist; even, initially, an anti-abolitionist. He believed (and in this was not alone among radical reformers) that efforts to free black slaves were detrimental to the plight of the working class in England, who were themselves enslaved by a feudal tyranny of land ownership. He claimed to detest the slave trade itself, 'knowing well as I do, that whatever the vile miscreants [the slave owners] wring from the carcasses of slaves abroad, they use for the purposes of making us slaves at home.' He even felt black slaves were better off than white English workers, if they were well fed and treated. These views, made popular among the working classes in his *Political Register*, were nothing short of a tragedy. A missed opportunity to align and link the causes of the two, and perhaps bring about change, understanding and reparation much earlier for the more critical, lasting plight and exploitation of the victims of slavery and racism. I would have liked to meet him.

Tenant, Tied

Past the halfway point of my husband Martin's tour and our conversations were full of the future, as well as what we'd do with the few weeks of freedom he'd be officially granted before returning to work. We both knew that when he *did* return to work, it would be relentless and all-consuming: that there would be little time for family or outings or to do any of the things we'd dreamt and written about. That working with horses (or animals in general) is a vocation, a way of life where normal rules of employment, especially when your house comes with your job, just don't apply.

One strange afternoon that spring, I climbed Beacon Hill with the dog and Billy in his back carrier. The walk up is a pilgrimage to a view. Once you get to the top, there is nowhere else to go but around the ramparts for the 360-degree panorama — and the knee-trembling slide back down again. It is not part of any way or trail; it is a diversion — it stands alone, worthy of a singular reason just to climb it. At some point, you are likely to touch the hill with the palms of both hands — whether to aid your climb up, to steady yourself on the descent or to support yourself, leaning breathlessly back in the grass. When you do, it feels like an affirmation; an acknowledgement of an ancient linear connection with people and this place. People are always stomping up, tearing down, hanging on to each other, slipping over, laughing. Beacon Hill has been loved and revered since forever. On the steep, soil-rolled slope above Ivory Farm, sheep walks and footpaths make scars and lines like stretchmarks and wrinkles on an old body. But the view to the west — fearfully, defiantly, possessively — felt like mine alone.

I walked past the great gateway to the south, where the slope is kinder, longer — and privately owned — and imagined drovers approaching: the travellers, the barterers, the tribes and their animals. I walked clockwise to stand by the wire fence looking down over the stud and our home. I could see the washing drying on the line, the yearlings cantering in a little gang below. The atmosphere was strange and unsettling, the air filled with a dulled, metallic haziness. I'd noticed a strange dust on the wheelie bin lid when I'd towed it back up the lane and, in the house, had caught the end of the news: sandstorms whipped up over the Sahara had been taken up and on by the jet stream, scattering shortwave blue-violet light and leaving us peering murkily through a long-wave filter of red and orange and sparkling silica.

On the hill, through an uneasy, autumnal, ochre haze, everything was tinged, foxed and coloured like an old map of itself – like those old sepia photographs of Egypt on the desk in the library at Highclere. Like my husband and Ondaatje's descriptions.

The weather got stranger; the clouds took on an unusual quality. The sepia air darkened and the sun turned a neon, streetlight reddish-orange. There were 'rain gods' walking the downs – those great, broad brushstrokes of cloudbursts that finger down from the sky and walk, obliterating their portion of the view in a slow instant. Great summer-fat raindrops fell from no clouds: a 'fox's wedding'.

I saw the first swallows come in, tail streamers snipping through the veil of fine, falling sand. Soon they would be building in the stables, swooping low over the heads of the horses looking over their half-doorways, making the yearlings start and shy for the first few days, until they grew forever used to them. I imagined them, and all the other migrant birds coming home from Africa, Spain and France, having crossed the Straits of Gibraltar and perhaps been strafed by gunfire. I imagined them up there as if in slow motion,

Tenant, Tied

rowing through thick, choking gold dust, on just an instinct. And I thought of him, too; and the sandstorms in another desert. I could taste it on my tongue, feel it drying on my lips, could stroke it to a fine silk powder on the inside of my wrist.

But there was something else in the wind, too. A palpable sense of time slowing almost to a standstill, before a moment of great change. The calm before the wind before the rain. The sudden silence of a flock of chattering fieldfare before they take off in a March prelude to flying north again for the spring; a sense of something approaching. I needed to go home, but I couldn't tear myself away from this foxed-photograph scene. I felt overwhelmed by the thought of Martin being so far away, in his own desert sandstorm, but also of the birds that were flying through this one to get here. I thought of the cuckoo that would soon be calling in the old bluebell woods where the Estate workers, who did the same job as my husband, would have walked the Church Path on Sundays and had last walks or picnics with their loved ones before going off to war. How must it have felt to those left behind, through all the springs that followed, to hear the cuckoo return all that distance, but not the ones they loved? I was rooted to the spot with the emotion of it all. Thinking too, that now, it was no longer a given that a cuckoo would return to these woods, their numbers being so diminished. Pierced on a barley twist barb of the fence in front of me was a large cockchafer beetle – the work of a shrike – a scarab beetle pinned like a Victorian specimen high above the land below. As I looked down, the haze thickened and my vision became grainy. Clouds of cottony fluff from distant poplar trees drifted up in clumps and across the landscape. I wanted to blink them away, but like migraine spots they seemed to move with my vision. It seemed as if everything conspired to obscure my view. I concentrated on the reassuring solidity of the mares and foals, putting our last conversation, about the possibility of ever leaving here, firmly out of my mind – when suddenly, as if instructed by some invisible ringmaster, the horses bowed their heads in unison and swung their quarters to the south. Rain and hail descended in a white arc, obliterating horses, house and hill from view entirely.

CHAPTER EIGHT

Home of Homes, A High, Clear Place

Exactly nine months and two weeks after my husband returned home from Iraq, our daughter Evie was born. She was a joy: bonny, curly-haired, overdue and with a wistful look in her eyes that, in our early days together, I took for disappointment. Her skin – unlike mine or our son's – was golden, as if it had the desert sand spangled in it.

But we were leaving Highclere. Martin wanted a different vocation and more time with a growing family than this half-a-day off a week job allowed for. He applied to train as an Emergency Care Assistant as a first step to becoming a paramedic and, from the moment he handed in his notice, we had just under four weeks to quit the house. Our single income (my job with a conservation charity became unfeasible when we had our son) would be the same – except there'd be no house with the job. Most of the rental properties we looked at were out of reach. But halfway through the third week of looking – and into my 41st week of an uncomfortable, nauseous pregnancy – we found a house on a near-neighbouring farm estate, an afternoon's walk along the ridgeway, in a village I knew well. Though the rent was below average, it would still take just under two-thirds of our income. I didn't want to leave Highclere. Billy was overwrought, confused and upset at the steady emptying of the house. I'd thought I'd raise him here, with his daddy and the horses and then a sibling, due any day. I clung to the idea of the new house warily, anxiously, as if it were a vine to cross a river that had broken its banks.

May into June – and a hard time to leave anywhere in England. Wild cherry blossom ringed the woods above the house and snowed confetti down from a high place. Cow parsley frothed the lane and elderflower plates bobbed, lit, pale and perfumed in the night. For days I had been unable to look beyond that – I half-pulled the kitchen blind on the hill and kept my eyes down when I left the house, avoiding the wood and the walk up to Heaven's Gate. Two weeks overdue, I went to bed with a serious migraine that my dear mum-in-law nursed. I'd never had a migraine before. I woke in the night and flung open the windows, convinced I could see a lamp swinging, beckoning me onto the hill.

The following afternoon, I went into hospital to be induced. Like one of the brood mares, I was heavy, fretful and resolutely not going into labour until I knew I was somewhere safe, somewhere that was home. In the end, labour was long, painful and difficult, the hospital busy and stretched – and I felt, as is often the case in childbirth, as close to dying as I did to a fierce, strong new life. There was no pain relief available and Martin had been sent home, only just returning in time. I tore badly, my pelvis was wrecked and I was stitched up roughly (still no pain relief), but it seems churlish or worse to mention that, when there she was: this little scrunched-up, reluctant swimmer-onto-dry-land. A daughter, safely delivered.

Home of Homes, A High, Clear Place

Travelling back from hospital the same day, verges bursting with wildflowers while insects appeared hyperreal, we seemed to pass everything in slow motion. I could see the light through blades of grass, intense shadow where they overlapped, the heads of individual flowering grasses – cocksfoot, timothy, fescues – all in sharp, lucid high-definition. I saw a bee's wing like a latticed window and nectar, syrupy and dripping from the heraldic tongues of honeysuckle blooms. Swallows and house martins dipped low over the fields for insects. As we turned into the long drive under the shadow of the hill, one of the Estate's tractors came towards us. We slowed and I raised my hand in triumphant greeting, but the driver didn't know us. I was horrified to see that he was mowing the verge with the flail: the flying chains mashing orchids, the yellow spikes of agrimony, bedstraws, the creamy plumes of meadowsweet, purple vetches, campions and scabious. All were reduced to a thick, wet mulch, a brutal pot-pourri, with horrible, indifferent, mechanical violence.

The tractor man that I'd expected – the charmingly named Tom Bowler – was leaving the Estate to retire. In his seventies, he'd worked here since he was a boy. He would not have been mowing the flower-rich verges at this time of year. He knew when to top the thatch so the cowslips could come for the bees, where the different orchids grew in different seasons – the twayblades and the burnt tips – when not to cut the herbal leys and where each mist pocket, spring or dry ridge lay. I felt suddenly afraid for the place without Tom's stewardship and wondered who might know or care what happened to this delicate, symbiotic balance. I felt a presentiment, a portent of unease; some change coming that I couldn't put my finger on. Hours after getting home, I could still smell the fresh mulch of flowers.

In our little bedroom, I fed the baby and, with my son, we watched house martins flutter up to their nests under the eaves, making that funny little '*beep beep*' sound of wet-blown raspberries my son could mimic so well. As we watched, a lone martin entered through the window, flew once round the room and fluttered onto a picture frame above the small fireplace. I wondered if it was last year's bird, returned to raise its own brood already – or if it was a more seasoned traveller, one that rowed through the thick sandstorm, just over a year ago, to come home. There hadn't been a thought then of moving, or another baby. Only a safe homecoming. The bird flew out and my son clapped, sending four sets of house martin parents diving out from under the roof like a red arrow display.

Twenty-four hours later, and with just five days left in our old home, I began shivering violently with a sudden high fever that developed into delirium and a serious womb infection – my damaged pelvis meant it was difficult and painful to stand, walk or bear any weight. My lovely mum stayed on as the rest of the house was packed up. On the day we moved I had to be half-carried out. The beech trees were in full summer leaf and the sunlight filtering down into the Warren gave it an underwatery, Pre-Raphaelite quality, like light through stained glass. My senses were still heightened. There was the faint smell of hay and horse and, where a horsebox lorry had clipped an elder tree, the distinct greenwood smell of bacon. Ghostly-pale points of light glimmered from the white helleborine orchids I'd only discovered the fortnight before, among the trees. They seemed to float before my eyes. I drank it all in one last time, felt the big wood at the back of my neck as a huge sacred presence. I insisted on locking the door for the last time, my throat closing with the effort

Home of Homes, A High, Clear Place

of keeping the tears down. I turned the big old key in the shoebox-sized lock and slipped it into my pocket for keeps. Nobody stopped me.

———

Highclere was my 'home of homes', a phrase that resonated from a poem of John Clare's when he left his first home at Helpston in rural Northamptonshire at the age of 38. Clare's poems were a detailed and intimate study of his very local farmed landscape, the wild familiars within it and his – and his community's – connection with it. Clare was a field-working, complex man, whose reading, writing and education (largely self-taught) set him somewhat outside the family and community circle. His real, hard-worked connection with the land, memory and its open spaces was unquestionable. He was utterly undone by the advent of enclosure which came late, but so brutally, to his parish. Thirty years younger than William Cobbett – whom he admired to an extent – Clare witnessed the iniquity of enclosure and the rise of the big, private game estates; yet unlike the Gentleman traveller and politician, Clare suffered directly under these seismic changes. Open spaces were measured, parcelled up and fenced off. Paths walked by timeless generations were stopped up. Common land, woods and water were denied and often ruthlessly patrolled to punish the new trespassers or poachers. Meeting places, folk history, traditional celebrations on the rural calendar, stories set in the landscape and even holidays such as Plough Monday were all denied, stopped up and lost, too, disrupting the flow, rhythm, life and relationship of things. Clare felt these were direct offences against community, nature, freedom and custom.

What must that have been like, this curtailing and clipping, this refusal, prevention and snatching away of rights and living? Of freedom? In Clare's poems, I had felt an echo; a whiff of what it had been like and meant, particularly at Greenham and at Newbury. Clare openly trespassed. His poems were less a romantic view of nostalgic pastoral scenes and more an intimate portrait of grief, loss, meaning and memory of a world he was losing. They were often a potent, outspoken and angry form of political activism against what was happening. Yet Clare was often a conflicted man. Divided in his views over the Swing Riots, he denounced a local incident of rick-burning that extended to the razing of a whole farm – and the horrible deaths of livestock. Clare was poor all his life, never owning anything much (least of all a house), and the big estate owners and landowners, employing the 'rogue' gamekeepers, developing their sporting estates and fencing off what had been community land, were his patrons, employers and landlords. Rereading his poems I felt the puff of a whisper of kinship.

A year before we moved from Highclere, I'd responded to a 'clarion call' from the author Richard Mabey at *BBC Wildlife Magazine* to write about nature, loss and the state it was in. The 'new nature writing' was in its infancy. I entered, writing about Greenham Common and my baby son – how I would make sure nature was a big part of our lives together, as well as his future – and won its 'Nature Writer of the Year Award'. Galvanised and encouraged, I began to write in earnest whenever I could, about the wild and farmed landscape we were immersed in, our relationship with it and the loss, always the loss, of biodiversity that resulted from the management of it on an intimate, yet industrial, local scale. I started a weekly column in the local newspaper, with

Home of Homes, A High, Clear Place

warnings from the editor, fellow Estate workers and tenants, and my family, not to be too contentious or upset anybody, least of all our own landlords and employers.

Clare was a vernacular, deeply political poet, yet not quite a radical. He was more a naturalist, upset and disenfranchised by 'each little tyrant with his little [private, keep out] sign' who, almost overnight (through enclosing open-accessed common land), erased and fenced out a means of living, many rural traditions and generations of connection with land and place for most people.

Clare's cottage and life in Helpston, working on the farms and great estates as a thresher, ploughboy, groom or gardener, would have been very familiar to my paternal great-grandparents – as would some aspects of life at Highclere. My grandmother was born (before her parents married) to itinerant Northamptonshire agricultural labourers. At some point, she was sent away to live with her grandparents, as the family could not afford to keep her after a sister and a brother came along.

In 1809, the year that enclosure came to Helpston, a 16-year-old Clare was prompted to write an angry lament for his 'dear native spot' and against 'accursed wealth, o'erbounding human laws'. In a poem about his village, Helpston, he writes:

Thou art the bar that keeps from being fed
And thine our loss of labour and of bread;
Thou art the cause that levels every tree
And woods bow down to clear a way for thee.

He repeated this sentiment in a poem called 'The Mores': 'Enclosure came and trampled on the grave/Of labour's rights and left the poor a slave'.

The act of enclosure marked the end of innocence for Clare too, the end of childhood. Both became a lost country he could never quite get back to.

In 1832, May into June (as I'd found, the hardest time to leave anywhere in England), Clare left the only home he'd ever had at Helpston with the encouragement of his patrons and supporters. They had secured a cottage on a neighbouring estate in Northborough. The house was a little larger than the tiny Helpston cottage, in better condition and with a cottage garden. It was a three-mile walk away. But Clare had misgivings. As Jonathan Bate puts it in his biography, moving away was 'a serious concern for a man who had derived his profoundest sense of personal identity from his immediate surroundings.' He felt strongly (and wrote) that it was as much that the place knew him, as he knew the place. An ancient, sustaining plum tree had blown down at the corner of the house and, poignantly, the last in a line of elm trees by the house was felled. 'All the old associations are going before me,' he wrote. That he had no influence over the felling of the 'old Elm that murmured in our chimney top' and 'rocked thee like a cradle to thy root' affected his already-uncertain sanity. Leaving his 'home of homes' added to a sense of unselving: the unhinging of a gate. Although it was a willing and legitimate (if coerced) move, Clare called his poem about leaving Helpston 'The Flitting', knowing full well all the connotations of this Northamptonshire phrase: to flit a house overnight, taking everything with you, to escape debtors or some other obligation. A 'moonlit flit' if you like.

In his time, Clare's poetry outsold that of Wordsworth, Coleridge and Keats. He visited London on a number of occasions and enjoyed the company, but felt the push and pull between home and away, country and town, labour and writing, never quite belonging anywhere. And with six surviving children (with his incredibly resourceful wife, Patty), he was always poor, to the extent that he worked for seven years on the hated enclosure gangs, blowing up trees with dynamite. Clare mingled with the literati of the day, but never met John Keats. He felt Keats and *his* Nightingale were too full of naiads and dryads and not enough true observation. Poignantly, on the fenland at Northborough, there were no nightingales. Clare left them behind at Helpston and, with them, became 'lost in a wilderness of listening leaves'.

The first few days and weeks in our new home on the Combe Estate was all windows and exposure. The former dairyman's cottage were one of six workers' cottages built in 1953 by John Astor MP on his estate, one across from Highclere and that had now been passed down to his son Richard. The cottage was a classic square 'council house' type with a flat, concrete porch. Four of them were occupied by staff. The house was on the northern edge of this Estate and the southernmost edge of a village, with a view of the downs and big, cold, Crittall windows. The front garden had steps down onto a narrow lane and stood between two farms. By most standards, we were out of the way, even a little remote, yet I felt exposed – as if everyone could see in. Someone calling at the front door could see right through to where I was feeding the baby. From the kitchen window, I could occasionally see people walking past the garden gate. I retreated upstairs. Whenever I looked out of the window towards the farm, I thought there was someone stood by the farm gate, watching the house. I had an overwhelming sense of having no right to be here that bordered on an uneasily lidded panic.

I was utterly shaken in a way I hadn't been by the birth of our first child. I had a newborn baby and a toddler in a new place, and my husband had just started a new job. I wanted to hide away with the children; the world seemed full of unnameable, unspecific and unspeakable threats and I didn't want to leave the house. I felt irrational and in shock.

I ventured into the back garden to watch a pair of red kites wheel in the blue sky and felt a sudden surge of joy bubble up like a balloon inside me. But then I caught the eye of my neighbour's teenage granddaughter staring at me, unspeaking through the hedge. I offered a 'hi' which went unanswered, and scurried back indoors.

I went out again to meet the kind estate manager, Chris, with baby Evie in my arms. He congratulated me on the birth of our daughter and welcomed us to our new home. All the time, my eyes were drawn to the vague shape of a person in the farmyard, staring. But, as we chatted about the kites and, reassuringly, people we both knew, I relaxed. Emboldened, I reassured him our dog wouldn't be a problem around sheep or in the shooting season and, as a tractor approached with its wide spraying arms out and running, in some perverse demonstration that I was a countrywoman for whom this was an everyday occurrence (and that we were therefore right for this place), I didn't step back. The sprayer came right up to the garden fence and, a concerned look on his face, Chris ushered me away from it: 'It's 100 percent safe, of course, but…the baby.' I felt suddenly foolish, alarmed and rattled. A fine mist of the spray reached us and, though I shielded Evie, I could taste it on my lips. The garden gate was wet with it. I made my excuses and hurried in, determined to bathe the insidious chemical off mine and my newborn baby's impossibly soft, absorbent skin. I dissolved into sobs before I could do it. My internal thermostat for danger seemed utterly skewed: unreliable and downright dangerous.

I felt unsteady, adrift and anxious; one step removed from everything, behind glass. I was impatient to know the view beyond the window – to connect – yet getting out there seemed insurmountable and fraught with wrong moves. Each time I made a mouse-like foray out, something sent me scuttling back in.

The first time I walked away from the new house took some coaxing from Martin and my lovely father-in-law. I felt immediately disorientated; frightened even. The sense of uprooting, of bare-rooted exposure, was more acute outside the house. I was afraid to cross the quiet, narrow lane. I was afraid of the cows in the field and that we might lose the dog – my normally reliable internal compass had gone haywire and I was afraid of getting lost. Where paths diverged, there were sarsen stones, placed as waymarkers. Even these touchstones loomed threateningly. It bothered me that I had no idea how old they were – or if they were the same stones marking paths we'd already been down.

But the walking began to help. The place was also full of promise and loveliness – open woods and commons and footpaths and bridleways in all directions, and all below the overarching magnificence of the chalk downland. I thought of John Clare's removal to Northborough, just up the road from his 'home of homes' at Helpston, which he felt knew *him*. At first, he was ill and couldn't write. Just a few years later, emotionally volatile, depressed and delusional, he was committed to a progressive asylum in Epping Forest. Four years later, he absconded and walked the 80 miles back to Northborough. Once here, he described feeling 'homeless at home'. How much did this have to do with his sense of place and belonging? With 'madness'? Or a sense that, as a reader and writer and studier of nature, he had set himself apart from others in his community and, indeed, his family, while at the same time intensifying a connection with his locality? Through solitude, wandering and deep consideration and connection with place he became more grounded, yet wilder. Clare was homeless at home, but also half gratified that he could be happy anywhere.

Cut loose, I took comfort from that. I had moved and moved, but the landscape was the same. With each move, I had only gone a handful of miles. I was from here. I began to allow the feeling that I was known here to surface. A feeling that the place knew me.

Returning home through the farmyard, I realised we were coming through the gate I could see from the bedroom window, below the big hill. I put my hand on the gatepost top, worn smooth by decades of farmhands turning through 180 degrees in the same motion, and wondered who it was that had stood here, looking at the house. The crossbrace and curved upright of the five-bar gate abutted onto the thick oak post. In the breeze, a thick swag of ivy blew across, partly obscuring it: here was my stranger at the gate. An illusion prompted by a disorientated and overactive imagination. Nothing more than a case of the shadowy, childhood dressing-gown-on-the-back-of-the-door.

Another foray into the garden. Under the shade of an incongruous, leaning eucalyptus tree, a shadow moved independently of the breeze and morphed into a hare. It was just feet away. It stood on all fours and arched its back like a cat, stretched and yawned, its long, leathery ears lying along its back like antlers before pricking up. I could see whiskers, a wild eye. It loped away across the field with an easy canter.

Home of Homes, A High, Clear Place

Again, that bubble of joy somewhere near my diaphragm. But at that moment, Tess the dog – who had come out behind me, stalking, unseen – leapt on something in the grass, threw back in her mouth something small and wetly furred. All too late I realised what it was – a tiny, newborn leveret. I yelled in horror, but she gulped it down in three mad-eyed, determined, uncomfortable swallows. Gone.

Things began to change incrementally from that moment. I was angry with the dog, with myself – and what I would have been able to witness *on my doorstep* had I been more engaged. So, I engaged.

I spoke to my new neighbour Betty, whose late husband had worked on the farm. As we gazed back at our houses, I noticed a grey, looped bunting across all four of our little row of cottages. The semi-detached ghosts of house martin nests against the fascia boards. According to Betty, a decorator had knocked them off ours and the rest had been broken off by an owl, nightly one summer, when it had found the chicks (despite Betty staying up to meet it with a broom). I thought of the house martin in the bedroom at Highclere and of how that bird – and other migrants like it – would know two very specific, local places in the world intimately. I made plans to put up boxes, make homes.

In the garden there were grass snakes and some old, friable owl pellets under an overhanging branch. And then one morning, a mole above ground. The shadows of buzzards rotated overhead like windmill blades, slicing at the sun. Pulling my sleeves over my hands to avoid a bite, I picked the mole up. Its black velvet, beanbag body was a warm, surprisingly heavy cylinder, with no visible ears and a short, bristly tail. The eyes were the tiniest pricks of light and its snout long, pink, very mobile and surrounded by stiff whiskers.

It wriggled strongly, a fine crumb of earth falling off suede fur that smoothed in any direction, allowing the animal to move forwards and backwards through close tunnels without resistance. I studied the mole's astonishing forepaws: huge spades negating any need for forelegs or arms. They were pale, human hands; each with five, long, white nails, the palms creased with lines a palmist could read. I showed my delighted son. But it wasn't until I put the mole down, away from the patrolling birds, that I recognised something quite anguished and animal in myself. Something so strong it felt like a revelation. Immediately the mole touched the earth, it went into action as if wound by clockwork, digging with incredible speed. Five strokes of its great paddle paws and it was gone, swimming into earth that closed over him as if he were never there. He was hidden, camouflaged, had dug in, gone native – gone between the roots that I needed to put down quickly or go mad. Quite out of the blue, this powerful instinct for going to earth struck me as a familiar and reassuring strategy. It was an act of possessing the ground and belonging to it utterly, regardless of who or what owned it. The mole had claimed the ground and used it as protection. It did not consider ownership at all. It felt like a good omen. It also struck me then, that I didn't need to 'come from' this place to be part of it. I only needed to get to know it, to let it under my skin on my own terms, wherever I was from, or 'coming from'. It occurred to me as a secret, rebellious thought that I carried around with me like a hot-water bottle up my jumper. Anyone could make a place their home by engaging with its nature. And that fierce core of a thought never cooled.

Above the house was the long whaleback of the down and the high dome of Walbury Hill. To the west and visible from the bedroom window was Inkpen Beacon; the only thing breaking the soaring silhouette of its near-perfect curve was the long barrow on the top and the double gallows on top of that. This was the same undulating ridge of chalk I'd left behind, just an afternoon's walk away, and the one I could see from my old teenage bedroom window – when I longed to be closer to it. If I climbed the first hill, I could see beyond the dome of Pilot Hill, the castle turrets, Heaven's Gate and wooded Siddown Warren. I felt the key in my apron pocket for reassurance, took it by its ribbon and hung it on the bedroom door. *Dulce Domum* (Sweetly at Home).

CHAPTER NINE

The Form of a Hare

I put down roots again, quickly, fiercely, strongly, to make it hard for anyone – or any reason – to move me next time. I fed the birds, borrowing the perky bright optimism of blue tits until I became like that too. I nurtured a brand of wilder domesticity, where the smaller things took up my time. At home (or rather outside) with the children, the world became more intimate, even as it became more infinite – worlds in acorn cups, heavens in a red kite's backlit wings, the thought-provoking questioning of children. The pace slowed, to allow the marvellous importance of a tussock moth caterpillar to cross the road, fully punked up and fully observed, or a muddy puddle to be completely unmapped, reconfigured and redrawn with a stick. I began to place things in the landscape, getting in through the children – and other wild things.

I saw brown hares occasionally in the young corn behind the house, but not again in the garden. I loved reading the weather coming in straight off the down and letting the washing blow in the big wind. It felt like a conversation – a kind of soft transaction I could have, if I read the weather right. If I didn't, of course, the washing would end up being hauled back inside, in armfuls of wet sails, to be heaped inadequately over doors and turned-off radiators, or, if I hadn't double or triple pegged it, spread over the field and hedge. With baby Evie in her Moses basket in the garden and Billy playing in the earth with the Britains Farm Land Rover that had been mine, I realised that the doe-eyed roe that appeared just beyond the garden fence mornings and evenings was leaving her own babies to lie up during the day. She watched us, cot sheets flapping on the line, like a living statue, before licking up her nose and leaving twin fawns in the long grass, as if in my care. Two new fawns for a swallowed leveret. A second chance to be trusted. I kept my distance, aware just of the flicker of a leaf-shaped ear, a sun-dappled coat tricking the eye in the waving grass. Life in the nursery.

Revisiting my own relationship with nature, I began to write in earnest to explain, explore, celebrate, mark, galvanise and commiserate it all. For me, the sombre threat of the pram in the hallway was never a problem – I was always out with it. Mostly, it set me free. I started writing a column for the RSPB's *Birds* magazine (that became *Nature's Home*).

Tiny hands led me back to the start of my relationship with the wild world; to explore new things and go over old with fresh eyes, a new wonder and the skewed luxury of time. Ever an unwieldy, tricksy thing for me, time spooled out at the beginning of days, only to clatter into itself irritably at the end with important, basic things left undone, a feeling of failure and a Victoria sandwich for dinner. My take on a chosen domesticity was jubilant, chaotic and spent mostly out of doors.

I set about homing, hefting, hearthing the children and *imprinting* them here. Not as an act as such, not with pressure, but as a zoological term I borrowed from the birds: those house martins, stable-barn swallows, nightingales and spotted flycatchers that return with absolute, unswerving fidelity to their natal and breeding site – their *nest* – each year until death. Travelling thousands of high miles in between on a season ticket, round or through storms, drought, exhaustion and the deadly reverse-hail of gunshot, they know those two most probably humble and unremarkable places in the world intimately: *here*, and *there*. I wanted the children to recognise, trust and habitualise themselves *here*. Home: their natural habitat. Landscape as surrogate parent. A base they could fly from and return to.

Under the eaves of the village hall, built on a ledge close against the warm brickwork, was the nest of a spotted flycatcher. Liking an open view, she sat, beady-eyed, tight to her round, cupped nest, like the lid on a sugar bowl. With freckled, mouse-grey breast, this long-distance migrant is a sweet, characterful and understated little bird. It has such a delicate, insectivorous bill, large eyes and gentle, enquiring expression, and there used to be one over every stand of insect-rich cow parsley which crowded round every stile or gateway.

Spotted flycatchers have simple, unfussy requirements: flying insects. But 60 years of unrelenting spraying, mowing and strimming in the countryside has dwindled their food supply to near nothing and is directly linked to their now-rapid decline. The children loved to watch her hunting techniques. Spotted flycatchers sit on regular branches and wait, looping out to catch flies with an audible 'snap' before returning to the same branch – a constant looping; a patient, productive, invisible crochet, a neat chain stitch of a flight pattern going back and back on itself. It is the act of 'homing' in miniature. A trip to the shops to snap up a bargain, and back. A trip to the shops and back.

The Form of a Hare

I paid close attention to passing on the human history of our new home through its wildlife, place names and stories. In such an ancient place as this (as in most of rural Britain) there is a rich, dense tapestry of interwoven, interdependent wild and human history. The warp and weft, the knit and purl of it, the stitch, the feltedness of it, impossible to pull apart without damaging one or the other, or both. I taught the children fieldcraft and weather lore, old wives' tales, birdsong, constellations and nonsense. I got to know people. The world was more alive to me and more meaningful than it had ever been before. Learning the names of things mattered – and didn't matter. I seesawed between the two, but the children taught me that mnemonic was everything. That its name didn't matter at all unless the thing resonated; was shown to be significant and meaningful – emotional, magical, funny, disgusting or weird. They made that perfectly clear. A thing's name didn't matter unless *it* mattered. Curiosity, discovery and connection became the key to memory and meaning. We made things up.

I taught them the 'teacher' song of spring and the great tit. Listen to that bird, I'd say, he is saying '*tea*cher, *tea*cher, *tea*cher'. In the pre-school playground, the biggest outrages are quelled with the words 'I'm telling'. So, the first bird Billy learnt in early January was the original 'tell-tale tit'. And by the time the spring chorus had swelled to a cacophony that included all the birds singing at once, Billy hadn't forgotten the great tit. Yellowhammers sang a picnic hallelujah. The '*little bit of bread and no cheeese*' my grandad taught me on the chalk of Portsdown Hill became 'pretty, pretty, pretty, pretty *pleeease*' can we have tea outside? And became the picnic bird. Even a bird that says its own name – that metronomic harbinger of spring – got corrupted. The chiffchaff became the chip-shop bird. Never mind vernacular names, let family names confuse the experience. I told the children about our 'treacle-birdie', my own family name for the blackbird and what he says on those warm summer evenings in such languid syrupy tones: a phrase that richly describes the bird's molasses-dark, glossy treaclyness, too. When I was 11 years old and living next to Newbury Racecourse at Greenham, the song thrush taunted the racecourse steward and the jockeys jostling at the start, calling from the top of the tall poplars, *'weighed in, weighed in!'* Or *'photo, photo!'*

Our litany does not stop at birds. There are antlebumps and standicastles (anthills and molehills) and the alchemy of trees whose names change with the season. The flowering of the 'ice-cream tree' is eagerly anticipated. The horse chestnut, resplendent with its towering white candelabras, becomes a tree bedecked with a thousand Mr Whippy ice creams, just when there are fêtes, May Days and parks with ice-cream vans. The children do not know what a candelabra is, but they know an ice-cream tree when they see one, especially the strawberry-flavour pink variety. And it seems a miracle to them that this wonderful tree transforms into 'the conker tree' in autumn with leaves like burnt, gloved, Guy-Fawkes hands that they shake, pull and run from, squealing as they are showered with the prickly green mines of conkers.

For a spell of time, night terrors would wake Billy and we'd blunder blindly into each other, our limbs still half-asleep, frightened by his screams in the velvet full-dark of the house. On these nights, with Martin working night shifts, I had to be the adult and quell my own fears. We comforted ourselves with owls. Opening the window onto the night and listening, we heard them call out, because (according to Billy) they couldn't see in the dark, and constantly asked one another, tremulously, half-afraid too: 'Who? Who are yo*ouu*?' And the returning 'kee-*wick*' from the female tawny became an 'It's me! You twit!' The scary night became funny, familiar and comforting.

In a place with virtually no road signs and where strangers are usually lost, place names were learnt, revived from old maps, or given new ones: spooky Pebble Arch, the Sticks Walk, Trappshill, Hog's Trow (for Trough, in the old vernacular), Rooksnest, Milking Parlour, Hell Corner Farm, the Pointless Stile (there was no fence). Later, when the children were old enough to be out on their own with the dog, on bikes, skateboards or the pony, this became very important. The field 'corner where we fall off' in particular, which had claimed an unusual amount of horse, pony and bike departures. I'd even managed to roll the pram there on one occasion, running after the pony with the dog lead inadvisedly round the wrist of my hand on the handle. The dog darted after a rabbit, pulling us all over and suspending the baby upside down from her harness.

I made regular pilgrimages to the reclaimed heathland of Greenham Common, where I could see the line of the downs. The children revelled in the wide-open space and the wind up there. Late April and May, we walked to the rusted 'fire plane' permanently crash-landed above its pit of brambles, wrens and adders. Billy loved to play around it and I came to have my heart freshly broken, listening to nightingales. He played, lost in his own world, while his little sister stirred cowpats with a stick, but the bird is hard to ignore with its suspenseful sense of the dramatic and is *so loud* – Billy and Evie would stop to peer at the scrub, looking for the 'fire plane bird'. It isn't ever given words to say. Its song is more than enough.

Naming wild things and places and giving birds lyrics for their songs, we absorbed nature into our own personal history and found a powerful sense of belonging and reassurance, connecting us with the place, the place in time and a feeling about it. It became part of who we are.

But this was no pastoral idyll either and no retreat. There were brutal reminders that it wasn't. I woke one morning with a jolt after a vivid 'bypass' dream had me walking barefoot through mud in my nightdress across what resembled a Civil War battlefield – only both Roundheads and Cavaliers were attacking a wood with chainsaws and the children were somewhere beyond it, calling, crying. I sat up, hair plastered over my wet face: Evie was crying and the sound of a chainsaw was not in my head – it was very close by. Evie was stood in her cot in a furious, frightened rage, arms out to me. I picked her up, her whole body racked with sobs, and opened the curtain. The ash tree in the garden that reached above the roof of the house was imprisoned in scaffolding. The two farmworkers looked sheepishly up at the window; one was on the platform and had already lopped limbs off the tree. I could feel shock and anger and downright indignation rising uncontrollably; I couldn't get the window open with Evie in my arms. How *dare* they? In *my garden*, without asking, without notice, without *knocking*? I flew down the stairs in my dressing-gown with the baby, Billy now awake and alarmed. I heard 'here we go' muttered as I opened the door, the chainsaw laying like a threat on the grass. Once, I'd held my arms out in front of running chainsaws and gangs of jostling men – these two affable and apologetic men in my front garden and a tree encased in scaffolding should have

The Form of a Hare

been a walkover. But I had a baby in my arms and it wasn't my garden. It wasn't my house. My home, but in no other sense was it mine. The then estate manager had the habit of calling our home or the Estate 'his': *my* woods, *my* field, etc. On coming in for an inspection one day, he asked me not to leave 'those clothes drying on the radiators, you'll damage my walls.' I had absolutely no rights or grounds to protest about the tree. I didn't own it. The men had the right to enter the garden of my home without asking, and chop it down.

I had become acquainted with the tree through four seasons. It was an airy trellis, creating a leafy shade over the pram in the garden and Evie's tiny bedroom, dappling the walls of the green-painted nursery. The luminous, feathery leaves were always moving, casting dancing shadows on the wall that proved better than a mobile. I'd known, when we made this Evie's bedroom, that in folklore and historically, newborn babies were believed to be under the protection of neighbouring ash trees, and I'd liked that. The health and medicinal properties of European ash have a very long historical association. They were already known in Hippocrates' time ($c.460$ BC); a spoonful of its sap was traditionally given to protect babies from ills and help them grow strong. There was even a superstition that the health of the two would then be intrinsically linked thereafter. I brushed that thought aside.

From the first-floor window, the open canopy was at eye level. We had a bird's-eye view into the heart of it. Blue tits performed among its twigs like acrobats. Tiny goldcrests searched, inch by inch, for insects. In autumn, the long, thin leaves covered the ground like little fish and, in winter, bullfinches raided the keys for seeds. A pair of wood pigeons cooed and coveted each other, cradled in its branches. Owls sat in it at night and called. I'd spent many hours diverted by the life of it, trying to soothe a baby who wouldn't settle – and kept myself sane. I've always loved how the branches and tips of ash twigs curve up gracefully, optimistically, to the sky at their ends with a flourish; and how, in recent weeks, the dusky lamp-black buds and smooth twigs looked like deer's hooves. Not more than six feet from the front door, the self-sown tree had broken and lifted ours and our neighbours' path, with a gnarly toe. It had become a tripping hazard, a safety issue. And no manner of pleading and explaining that I would paint the path section with luminous yellow paint could change what had been decided. Strange to think that at this time, although I didn't know it then, and it wasn't officially recognised until 2012, the first cases of Chalara Ash Dieback were being reported in the southeast of the country. A windborne fungal disease that European ash is susceptible to, a problem exacerbated by imported infected nursery stock, it will kill around 80 percent of our third most abundant tree. Whole woods will be lost. Landscapes changed forever.

I thought of John Clare and his Helpston Cottage Elm. I retreated into town for the day, my family comforting me with the promise of an apple tree to plant.

Returning home, the sky looked bigger, the house blank-faced. A thin trail of sawdust curved across the lawn. There was nothing left but a white stump and a familiar, unsettling smell of petrol and greenwood. I counted the rings carefully and was shocked to find the tree had been exactly my age: 35 rings of grain and a foot for every year.

No one ever came to repair the broken-path trip hazard. More than 15 years later, the treacherous bump and crack in the path are still there. No one has fallen over it yet, even on the darkest of nights. And a revolutionary line of pliable ash volunteers has sprung up in its place. Like the daughters of Boudicca, who released a hare from the folds of her dress and followed it onto a battlefield.

At some point, as we settled into the rhythm of our new life around the lurching Cake Walk that is shift working, we dared each other shyly to think about another child. Could we? Should we? Indecision, consequences, finances and responsibility battled it out like weather fronts with the thought of new levels of joy, chaos and the creative wonkiness a family of five might bring. We introduced chance into the mix, and were a bit hit-and-miss with the contraception. By the time we went on holiday, camping in Wales, I was pregnant half-unexpectedly and caught in a maelstrom of hormonal changes and all-day morning sickness. I was wildly moody: sharp-tongued, then tearful and remorseful. I worried that we might have been irresponsible, even done the wrong thing. That I wouldn't cope with three children under five, not enough money and a husband working twelve- (often fourteen-) hour shifts, day, night and weekends. I worried what my parents would think, what my in-laws would think – my friends and, of course, the children. I even worried about my beloved (very part-time) job at the library in town. I worried about not belonging here, in Wales, on holiday. My worrying bordered on the irrational.

We pitched our tent high above St Brides Bay, the temperature plummeting by half on our first day. The Pembrokeshire coastal path was stunning: rocky cliffs of green sea beet and yellow rock samphire, the white umbels of wild carrot and pink sea spurrey tumbling down to a sparkling, amethyst sea and isolated sandy coves full of warm tidal pools. There were tantalising glimpses of the seabird islands of Skomer, Grassholm, Skokholm and Ramsey. I desperately wanted to go, but the children were so small and the weather wasn't easy. On the beach, the wind took the empty buggy and sent it racing by itself down to the shore and tipped it into the sea, while hurling sand in our faces like a volley of abuse. Overnight the wind worsened, shaking the tent vigorously, so that it strained at its guy ropes. In the morning, any hope of a trip out to the islands was dashed along with the possibility of dolphins. There was a violent swell in the notorious Ramsey Sound.

We went rock-pooling instead at St Brides. There were deep, warm, womb-like marine pools surrounded by seaweedy rocks. Whelks cluttered together in crevices and beadlet anemones waved their dahlia tentacles underwater and retracted into sticky jelly-blobs out of it. Billy was enthralled by the rings worn into the rocks, 'home scars' made by limpets holding fast in storm tides and above the tideline. It fascinated me to think that an organism such as a limpet had evolved a homing instinct, key to its survival. I stirred the water. Transparent shrimp sculled like grainy images on an ultrasound – there and not there, until they quickened into existence again with movement. I forgot and remembered and forgot that I was pregnant again.

Then, on what turned out to be the last day of our holiday, the wind dropped and the sea calmed, just as I began to have misgivings about a trip out to Ramsey. This was a family holiday on a tight budget, and the wild weather meant it had cost more than we'd planned – cooking had been impossible. Just Billy and I would go, but, at nearly four, was he old enough to 'appreciate' such a trip and, selfishly, would I be able to enjoy this opportunity? We caught the boat bus from St Davids and boarded another boat from the lifeboat's slipway anyway. Billy was the youngest person by far.

We landed and he gabbled merrily through the ranger's talk, telling everyone that he was a 'Wildlife Explorer' and worrying that he'd lost his badge. He was transfixed by the plastic bird models above our heads. The circular trail was a steep, boulder-strewn 3½ miles, taking on average 2½ hours. We had just three hours on the island before the weather was due to come in again and I knew I could not carry him. But we risked the walk round. It was too beautiful and intoxicating not to. The cliffs were a bee-buzzing pink, yellow and purple. We spotted wheatears, wild ponies, five colours of rabbits and some of the highest sea cliffs in Wales. And choughs. I heard them calling their own names down the summit of Carn Ysgubor, careering down the mountain, red-beaked and glossy black. We watched them turn themselves inside out on long shining wings with an awesome aerial grace, double-somersaulting against the cliff face. Firebrands. With their flame-red bills and legs, they were thought to set fires in ricks and thatched roofs by stealing smouldering sticks and burying them in hay or straw.

And suddenly, we'd reached halfway, ahead of ourselves. We paused near the summit of Foel Fawr, looking out to sea at the Viking island of Skomer, scanning the currents for porpoise. There were seals in a small cove, but they were hard for Billy to see. They were copper-dappled or blue roan beneath the waves, bobbing like bottles in the water. I wished he could see what I could. Through the binoculars, I could gaze into their anxious selkie eyes, and imagine there were answers there – or reassurances. But what I saw was an anthropomorphism of deep concern and worry.

We reached the jetty with enough time to buy a new RSPB badge and a carton of juice. Wide-eyed and tired, Billy told his dad his favourite bit – the swallow's nest built on the rump of the plastic display raven.

That evening, an unforecast wind blew in from the Atlantic, putting an end to our holiday. The campsite flapped and rang with mallets into the night. We got up to help others secure whipping guy ropes. One couple watched their small two-person tent tear right open. At 2.00 a.m. our tent turned itself inside out, tipping the children out of bed and snapping its carbon-fibre poles. It belled and billowed in and out, like a tethered jellyfish, inverting to press on our faces, then ballooning up and almost away. In torchlight swinging wildly from the broken crosspoles, we passed the children out to the car, frightened, crying and wet. We lay awake as the storm rocked the car and watched the lighthouse beam smooth the sea from St Ann's. I tried to imagine myself with three children. I would not have enough hands to hang on to them all. I slept in snatches, dreaming fitfully of trying to hold on to kites and, one by one, letting them slip. Near dawn we all woke, cold and uncomfortable, but safe; blowing on a warm glow of adventure. The wind had dropped, the dark sky was scoured clean and we watched as the first of the Perseid meteors pitched and fell out of space, bright sparks of comet dust spinning through the sky. A ritual we've not missed since.

Three weeks later, shelving books quietly in the library, I experienced sharp, low pain and found that I was bleeding. Two days later, the grainy, graphite image of my 12-week scan revealed an 11-week-old baby, but no heartbeat. My womb, like a rock pool, swished gently round the tiny image of a nestling chick, fallen too early out of its nest and gone to the bottom.

It was a shock then, to realise how much I wanted it, with a crushing, guilty grief.

We would have had the loving support of friends and family.

We would have managed. We were in receipt of housing support. Things would have got easier.

But, like a constant thumb over a worry stone, I'd worn and fretted her away.

When I was well enough, I walked up the hill to lie by myself in the chalk grass. There were lapwings. I was close enough to hear the creaking beat of their owl-like wings as they dived, tumbled and called. The soaring *'pewit, wit, wit- eeze wit'* that followed was quite the loveliest, most joyful sound I could want to hear. Blue-black clouds dark as swallow wings gathered in piles above the wood. I lay back among the harebells, clustered bellflowers and the little purple hop-hearts of quaking grass, shifted to avoid a thistle and disappeared. I scrunched my eyes and mouth tight, as a harvestman spider ran over my face. Burnet moths collided with tiny grizzled skippers in the air above me. A buzzard mewed. Out here, at this moment, nature decides what happens. Everything connects and has its place, its lifespan, its small death; every thistle its bee.

After a while, I glanced to my right and spotted, some 15 feet away, an eye – a glowing gold coin on its edge in the grass. It belonged to a brown hare, crouched almost beside me, long ears lying along her back like sundae spoons laid down on a table. How long had we been sat beside one another?

I watched her for a while before I knew I had to get back to the list on the kitchen table and the children, minded by a neighbour. I got up slowly, stiffly, and her ears rose smoothly, mirroring my movement, like the crest on a bird's head. Then she was up on her haunches and away into an easy rocking-horse canter.

There were two indentations in the grass – hers and mine. I laid the back of my hand in her hollow empty form to feel the still-warm grass.

CHAPTER TEN

Flint, Feather and Bone

A year or more later, a second daughter, our third child, was nearly born like a lamb into the mud and straw of the Ford Escort's footwell, beneath the blades of the giant wind turbine on the M4's hard shoulder. Thank goodness she wasn't. My husband had the skills, certainly; was calm and had already delivered several babies who wouldn't wait for a midwife, but 'if we could make the hospital, it would be better'. She was born within ten minutes of our arrival at the Royal Berks, in Reading (situated, significantly I felt, behind The Museum of English Rural Life), into frantic silence, a collective holding of breath and much urgent action among fantastic NHS staff, before being whisked away. Again, everything seemed to slow. She was placed briefly into my arms and gone again for checks. But she was fine. She was absolutely fine: unbeknownst to anybody, at some point, she had swum around in my womb and through a loop she created with her own umbilical cord. She was born with a 'true knot' in the cord that had pulled tight as she was being delivered. A true lover's knot. Things could so simply, with the pulling tight of a cord, have ended very differently. I've never felt so grateful for anything in my life before.

A few weeks later, with none of the ill effects of last time, I was out of the house, pushing the pram with our new baby in it up the hill: mum to three children under six. The cold early November air was exhilarating and we walked into a countryside on fire. The rich buttery golds of field maple, the long red leaves of wild cherry, and the unexpected reds and purples from dogwood and elder in the hedges were hearthed and crackled by the first fallen beechmast and the black marks left on the road by long stalks of ash leaves before they rolled into the roadside, their leaves stranded like little green fish.

I pushed up the old holloway and through a wrap-around tunnelled aisle of abundance: seeds, nuts and berries, hips, haws, damson and bullace, strung with necklaces of wild clematis, black bryony, nightshade and the little green cones of hops. Ivy flowers smelt warmly of honey, lemon and winter hay above the nostalgic old-book smell of tar spot fungus on sycamore leaves. Burdock and agrimony burrs velcroed themselves to the pram hood and the sleeves of my coat, like tiny, hooked planets.

We broke out at the top of the tunnel and bumped along the flint track to the beech hanger, the dog running rings around us, ears flying. The view over several counties was a joy. A whinchat on its autumn migration gripped the wire fence along the old drove road, long claws overlapping like a butcher's fist of tiny knives, mimicking the barley twist barbs it sat between. Then, in a presentiment of rain, a breeze picked up, rolling the brown-and-silver cigarillos of whitebeam leaves along behind us, making me turn to see what the dry, ticketty sound was: autumn and time chasing our heels, like low, nippy collies.

I'd walked too far in my eagerness and knew I'd be pushed to get to pre-school and then school on time *and* feed the baby. But for that moment, on the hill I'd sought many times before, I felt free and happy and high above it all. Capable of anything. The breeze came again and rustled the tops of the beeches. In a glorious rush of sound and colour, the leaves came down like ticker-tape. A confetti of toffee pennies that fell on my face and the baby's pram, welcoming her into the world. I laughed out loud. Above us, a kite tilted on its tail.

I felt fêted.

I lived my weeks within a mostly slow, child-paced, walking distance of home. We'd bought a second car when we'd first moved here, so I could still get about – to playgroup in the next village, to shops in town (there were none here), to my part-time job at the library and to see friends; but now, one car had become unroadworthy and too expensive to fix, and we couldn't all get in the other, as it only had four seat belts. Cars with five seat belts were newer and came at a premium. My world shrank. For almost a year I had no car, unless Martin was off work or sleeping off a night shift. I had to be resourceful and try to plan ahead, which was fraught with problems with him on almost permanent 'relief' shifts, without a discernible pattern and subject to change. A standard twelve-hour shift rarely finished on time and regularly became a thirteen- or fourteen-hour one, with travelling on top of that. There was a bus service, but it was difficult to fit a trip to town and back between pre-school and school times. Yet the kindnesses shown, the everyday offers of support and help from neighbours and friends in the small community were overwhelming. I learnt to see them as genuine and not be afraid to accept them – and reciprocated whenever I could.

I borrowed, helped with or looked after ponies we couldn't afford ourselves. There was a rocking-horse dappled grey called Lucy at a big house I cleaned for and a Welsh

Mountain pony, named Tiny. We taught Evie to ride and made quite a procession round the village with the dog, Evie up on the pony, Billy on his bike and baby Rosie, snug but cumbersome in her papoose. It was tricky when the going was slippery and I couldn't see my feet. There was nowhere to put my binoculars and, of course, I was desperately compromised if anything went wrong. But there were moments in the last snatched hours of a winter after-school weekday when the white pony stopped to watch a white owl, alerting us to it through her pricked ears. The moth-light, silent quartering of the barn owl became a regular feature of our week. Another time, at the peak of Billy's fascination with birds of prey, we saw a peregrine falcon. Only, it wasn't, it was a wood pigeon, barrelling over our heads in slim, grey, lithe panic and jinking away when it saw us. But he was so utterly convinced: 'Peregrine *falcon*, look!' Who was I to disappoint him and his toilet-roll binoculars? His sighting was so very well earned – and who hasn't occasionally mistaken a wood pigeon in its soft navy serge and RAF colours for something far more exciting?

We made sense of our landscape, investigated it in forensic detail, walked everywhere. I met the gamekeeper on the Estate on a school 'countryside day' where we collected basketfuls of pheasant eggs in colours like a Farrow and Ball paint chart. He patiently showed the children the trick with the egg candler, a light stick held to the egg to illuminate the blood vessels and developing embryo inside a fertilised egg. But the best part of the day for us all was taking the full-term eggs out of the incubator and letting them hatch into our cupped hands, so that we each got to hold a tiny wet creature until it became a fluffy, stripy, cheeping chick, impossible to fit back into the egg halves it had just come out of. We explored a huge, ancestral, active badger sett in the woods behind our house. These were impressive, ancient earthworks. Great chalk ramps and screes tumbling down embankments where knobbly, clawed-out flints had settled against tree roots and paths were worn hard and smooth as pavements. Embedded in the excavations, the gamekeeper showed me plastic ICI fertiliser sacks, their vintage logos still bright from another century. Some 40 or more years old, they had been stuffed with earth and rubble and used to stop up the holes before the setts were gassed and any badgers in the sett at the time killed. Gradually, though, the sett had built up again and this one included several families of badgers. Billy came home with a lump of chalk scored deeply by bear-like claws and, the biggest prize of all, a badger's skull, missing the lower jaw, kicked out of its final resting place when the sett was needed again, the long-buried body long ago decomposed.

The primary school's emblem is a child's (red) toy kite. The gamekeeper used the opportunity to explain that it was named after the real thing circling above us, while dispelling a few myths about 'the threat and damage red kites do' quicker than I could. He quietly mentioned he and his family had been part of a re-introduction programme here, under the jurisdiction of the then English Nature. When I expressed my surprise, he confessed that while keepering was his job, conservation was his passion. We hit it off there and then.

I returned to the badger sett that spring, often. And, against a soundtrack of young buzzards calling for a last evening meal and tawny owlets calling for their first, I watched badger cubs come out and play. A big sow badger first; eye stripes brightened by chalk powder, having the same effect a handler might employ to whiten the socks or tails of cattle for the show ring. She scratched loudly behind her ears with a hindfoot, before shaking her whole body like a wet dog, sending clouds of chalk dust out to settle on spent bluebell leaves.

I went out with the 'keeper too and we once played Grandmother's footsteps with a sleeping badger cub until we were close enough to see how the powerful digging claws were clogged with a marl of chalky earth and how fine the blend of black, grey and silvery hairs in its coat was: a camouflage for moonlight. The cub woke, uncurled and turned briefly to face us and the moon. There was a point of light on his wet nose.

In the spring before Billy was eight, I took him along. We took care to wear dark, 'quiet' clothes. The wind was up and, although not entirely in our favour or faces, the rustle it created helped mask any wrong moves or fidgets on our behalf.

Nearing the place, we cleared our throats, blew our noses, dropped our voices and practised rolling our footsteps, heel to toe along their outside edge. Then, I pulled out my trump card, easing the wrapper off a lolly in my pocket with my thumb and handing it to Billy, who, as planned, fell silent. We tried to creep in, over last winter's leaf litter, loud as crisp packets, and sidled through nettles taller than he was, settling to wait against the rough bark of a big oak.

Then, through nettle stems, white stripes. A train of badger cubs emerged cautiously and then in reckless bursts, just metres away. They canonned into each other and began to bite each other's rump, tail or ears, causing the one in front to bounce round and face the others. The last two stopped to nibble and 'flea' each other between the shoulder blades until one tucked its nose in and tried a slow, comical, forward roll. We stifled giggles. Stuck against a root, its bottom was in the air, so its sibling nibbled at that and continued to nibble in the same place (now its soft belly) as the upside-down badger slowly collapsed onto its back. There was a yap and a brief, whickering fight until a blackbird's shrill alarm drew all our attention, the cubs having also learnt the rudiments of a language of another species. When we looked back, they had gone.

A muntjac barked at 11-second intervals and we mouthed counting them soundlessly, a slow metronome to an early summer's evening. Then, as if from nowhere, badgers from a different direction: four pairs of tiny crescent moons for ears rode the broad planets of their skulls down the path towards us, three strides away.

They passed by, battle-ready claws rattling like sabres against the hard-baked path, the soft lozenges of their bodies following on like cloaks of moonlight. Then Billy gave a loud, accidental suck on his lollipop. The last badger stopped and looked straight at us, *through* us, moving its head up, down and sideways before bolting and setting its siblings off in a gallop into the dusk. I thought for a moment Billy might cry, having broken this magical spell. But we exhaled deeply before creeping out of the wood, grinning from ear to ear. I looked back at Billy to see that his favourite hoodie had come into its own, its glow-in-the-dark mask of Darth Vader appeared to be floating. It looked badger-like. We were suddenly breathless with laughter and what the badger cub made of it. And then we were whooping, running across the darkening field together, scattering wood pigeons gone to roost.

The spring just before each of the children were eight turned out to be the optimum time to begin badger watching. We sat among them as the cubs played and tussled, got stuck in the fork of a low tree, artfully gathered fresh bedding and humped it backwards, reversing expertly down a tunnel without wing mirrors; we have watched them roll in patches of wild garlic to stave off flies and fleas and watched cubs fall asleep mid-play, their concave heads laid one on top of the other, like precariously piled dishes. On one evening as we left a sett, our path was lit by the luminescent glow coved deep in the cowl lanterns of cuckoo pint.

I started a Wildlife Club at the primary school, along with the children's teacher, Kate, a geographer and herself a Newbury Bypass veteran; she had once halted the progress of a JCB by standing in front of it with her one-month-old baby son, Luc. Together, we dissected barn owl pellets brought in by the gamekeeper from the nearby church, made black paper cutouts of bats, held worm-charming experiments and once, incredibly, found a sleeping dormouse, rolling across the playground in its gently unravelling nest.

At home, we curated our own Natural History Museum in the shed, with outlying and mobile collections in pockets, on windowsills in bedrooms and, often, in beds.

Up on the high chalk of the hills, cultivated fields of rattling flints proved fertile ground for fossil hunting, if not for wheat or barley, grown anyway on the tyre-hungry, broken mosaic of ancient, knobbly silica. Among pocketed feathers and eggshells, thumb-polished conkers and leaf skeletons, there were potential arrowheads and flint axes of those that had built or visited the hillfort. Once, there was a sheep's bell.

Near a plot of land ploughed and left for stone curlews, we found glow-worms one evening; little ethereal points of green LED light among the thistles and arable 'weeds' of disturbed ground: fragrant pineapple weed, fat hen, shepherd's purse, shepherd's weatherglass (scarlet pimpernel) and, for a while, the rare shepherd's needle. The fact that the fossil teeth, toes and claws of prehistoric animals we'd discovered turned out to be flint burrow infills (solidified liquid silica, making plaster casts of animal tunnels in the ancient seabed) did not diminish our enthusiasm. Sometimes, there'd be a sea sponge 'rattle ball': spherical flints wrapped in smooth chalk that had formed around a sea sponge, long since disintegrated. Round as cannonballs or dinosaur eggs, sometimes the size and weight of cricket balls, the hollow centres would rattle when shaken. Best of the fossil finds, however, were shepherd's crowns, fairy loaves or thunderstones. These were weighted with a lost kind of alchemic currency and significance. The domed or sometimes heart-shaped, chalk-and-flint pieces were stitched into fifths by neat parallel lines – paperweights imprinted with a star shape, a fairy's Kaiser bun. Knowing they were fossilised echinoids or sea urchins did little to dispel the reverence with which they were handled and replaced. Only occasionally did one come home – we'd schooled ourselves to leave most of our 'finds': a favourite is encased in its own saucer of sea-green flint, its edge, strop-sharp, its dome fitting exactly into the palm of a cupped hand to work it.

Some of our finds were of a more grisly nature: fox and badger skulls discovered near setts or old earths in winter when the undergrowth had died back, and a road-death roe deer skull, complete with pearlised antlers. Sometimes, under hedges or in the woods, we'd find the great, flat, tined palms of fallow deer antlers. Golden syrup tins were packed with moulted owl, kite, buzzard and raven feathers we'd found and there was a boiled sweet tin of tiny feathers – goldfinch, great-spotted and green woodpeckers, jay and woodcock. A tiny treecreeper's skull, thin as tissue paper, sat on a shelf next to the enormous anvil of a raven's skull. Both from found, dead birds, the decomposition process aided by the surprise arrival of an importantly tabarded, orange-and-black sexton beetle. Each time, I lifted the tender hosts of their corpses to a molehill, protecting them with a bird-feeder cage. We watched as this evolutionary wonder of a beetle began to recycle the dead and spread their nutrients through the earth, creating a nursery in each feathery 'crypt'. The chemoreceptors within its antennae may have detected our dead birds from up to two miles away. Each time, by the next morning, the carcasses had vanished. Marked with a stick, I was able to dig them up for the skull, months later.

The first dead animal to be left on my doorstep came with a text message: 'So sad, owl hit by car in front. It looked so perfect, I couldn't bear to leave it, thought you'd like it.' I took it into school. The children were fascinated, awed, keen to feel the depth and density of downy feathers and discover how they might trap air; the sharpness of its small, mouse-eating beak, its powerful, furred legs. They marvelled over its incredible sharp-curved talons, over the deadly hind claw that pierces its prey in an instant when the bird closes its toes – and the strange-shaped, spongy gripping pads of its feet.

I gently spread its wings like a fan, revealing the intricacy of flight feathers patterned with a broken line of white, and showed them how the feathers slide neatly back over one another to close. Nearly a metre across, they drew gasps of astonishment. One child exclaimed they'd 'never seen so many styles and colours of brown'. They didn't quite believe it was dead. And I suppose it had taken on a new life of its own; phoenix-like, it took flight in their imagination. I buried it in the field behind the house and marked its position, hoping to take the skull back in a couple of months to show the schoolchildren.

The next dead animal to arrive on the doorstep in a 'bag for life' was a polecat, killed on the road. Earlier in the year we'd accidentally caught one in a live rat-trap cage (part of our 'wildlife-friendly' arsenal to try and quell the booming number of rats in the garden). We'd been woken by bloodcurdling yowls and murderous wails and gone down in our dressing-gowns to find a furious dervish of a caged animal. Confronted with us, the animal initially froze, but it met our astonished gaze with intelligent, bright eyes from its racoon-like, bandit-masked face for a moment, then it bared cat-like teeth and long, sharp canines and yikkered loudly and fiercely. The children, vulnerable in bare feet and pyjamas, took a respectful step back. With thick gloves and a garden fork, we levered the cage open and it flowed out and away with a sinuous, sable vivacity and undulating, stoat-like bounds, its absurdly bushy tail marking it out as something totally different.

The body in the bag was fascinating. The animal's spine had been broken, but other than that, the body appeared undamaged. It was covered in ticks (and already beginning to stink) but otherwise seemed healthy. We explored its teeth, the intricacy of its foot pads and claws, and the rich, two-tone luxury of the animal's famed dark-chocolate-and-cream sable fur with respect and awe. We buried it, but, along with the owl, never found it again to preserve its bones.

My RSPB column expanded into writing and creating new family activities and I co-opted the children into exploring creative and practical ways to engage with 'nature' and 'outdoors'. The talented and instantly likable RSPB photographer, Andy Hay, came down four days a year to photograph us, out of kilter with the two-seasons-ahead deadline of a quarterly magazine, building bug homes while wearing hats and scarves on the hottest day of the year, or pond dipping on a day when we had to break the ice on the village pond – in our summer T-shirts.

Long before we knew it, nature – or outdoors – became a common ground, a neutral place outside of home, but where we felt at and found a home ground. A base from which to grow from and return to, a place of freedom and of emotional safety and expression. Perhaps most surprisingly, this outdoors home ground became a place from which to explore new experiences and mediate change – 'outdoors' came with us, wherever (almost) we were. A constant colloquy between ourselves – and us and the place. I realised that to a greater extent, 'home' wasn't the house at all, of course. It was its footprint, its exact location, spreading from under its skirts and away, that we'd come to love. I'd loved all my houses as homes, each one. Dragged my feet when we'd had to leave. But what I remembered in sensory, mnemonic detail, was *where* they were and what surrounded them. What fresh smells, where the prevailing wind came from, what birdsong, what trees, what paths away. 'House' was just a means to live there. Shelter, warmth, sustenance; a stable in a bigger field.

The ground around our home became our ally in the negotiations, joys, frustrations, celebrations and trials of family life. One such negotiation happened over homework. The arbitration that goes on between a school that must conform to a model, expectations and national standards, compromised, floundering parents and a child that sometimes couldn't or wouldn't learn in the prescribed way, caused a fair amount of tension, anxiety and stress at times. The walk to school alleviated this at both ends of the day, taking us through woods and farm fields where a pair of kites nested or roosted in an old boundary oak, about 650 feet from the house. Their gentle plying and tacking on raggedly elegant wings became a daily fixture. One morning, when we emerged onto the road, one of the birds took off from the ground just ahead, its wingspan of nearly 6½ feet seeming to take up the whole of the narrow lane. It had swooped at something on the road. We took a few steps back and watched it come in again, low over our heads, to swipe at a small pale mound on the tarmac. Each feather was visible, its colours vibrant. It rowed away and we walked on to find its prey was a custard cream; mine and my son's all-time favourite biscuit, fallen out of someone's lunchbox and crushed by a car tyre.

Later that week, we watched the kite through the kitchen window with an after-school cup of tea and half a packet of biscuits. Billy was in his penultimate year of primary school. I had to remind him again that it really was time to get on with his homework. He'd stalled and eventually stomped off with a slam of his bedroom door that shook the window in its frame. We were both at the end of our tether and our frustration escalated into a row.

Minutes later, the bird banked in low and swept into a tight spot between the hedge and the tractor shed before soaring up again, making small, slow adjustments to balance itself, like someone riding a bike at walking pace. The space didn't seem large enough to contain it. It came round again, level with my son's bedroom window and so close; I knew he had seen it from the loud exclamations that came from upstairs.

Three jackdaws mobilised, hopping off the oak's stag-head to dive-bomb it, but the kite came round again, chestnut, white and black primary feathers riffling, pale blue-grey head and hooked yellow beak out in front. It rose vertically with barely a wingbeat, sheering off with breathtaking agility to sweep again past the window. Above my head, to the right, I could feel my son holding his breath above a bubble of exhilaration that squeezed out his anger – just as I was doing. The kite continued to make constant, deft, minute corrections to its flight. At one point, its wings were held at a 45-degree angle to its delta-shaped tail: a rudder to the great sail of its wings, a counterplane.

Billy had a red kite flight feather as long as his forearm. He wanted to make it into a quill to write with and we had promised him a bottle of coloured ink, if he did his homework. He was sure something magical would happen if he had this quill to write with. What he didn't recognise, as such, was the alchemy that had already taken place. When I went up later, with hot chocolate and a custard cream as a peace offering, red kite had become 'thought fox'. The page was printed with Billy's spidery handwriting: the boy and the bird and his freshly sharpened pencil had created something. He'd effortlessly linked his spelling practice sentences together into a required story: King Arthur, a bird of prey on a 'gorntlit' glove, Merlin and a spell written with a quill from a phoenix. A bird risen from the ashes of graphite dust, temper and the tinder of curled sharpenings. In the moments sparked between a dozen wingbeats, he'd written two pages.

Ours was a small, close world and, at every opportunity, we tried to make up for any lack of cultural or social diversity, questioning any preconceptions with the children until they began to do so themselves. In some ways, there was a sheltering, and in some ways, a lack, but out in those big winds, sometimes on their own, they learnt how to bend and give. During the pain and mental anguish of GCSEs a few years later, Billy coped by belting up the hill on his bike in all weathers, until he could fly level with the kites, 'getting air' from the dips, ruts and hollows on the hard ridgeway track.

Our little flock had what we needed: shepherd's purse, bell, needle; weatherglass and crown. Flint, feather and bone.

CHAPTER ELEVEN

On Gallows Down

When we moved to the village of Inkpen, there was still plenty of evidence of an older, poorer, more isolated and rural existence. Several cottages and farms were in tumbledown or derelict states, inhabited or abandoned.

Our 'unnamed' road went by different names on several maps – each end was named, but not our bit in the middle, so it was variously Rooksnest, Hollow Lane, Great Common or Wergs Lane but was between them all; the latter shared its name with an ancient, black, oak-raftered Berkshire barn that we watched fall into fascinating and romantic ruin, haunted by barn owls. Road names and directional signs were few and far between. Notices of road closures (due to frequent and persistent water leaks) were a conundrum of intersecting 'unnamed roads'. The village reputedly had the dubious title of 'most spread out village' in Berkshire with 'the windiest road' and was really several small hamlets separated by woods and fields, built around ancient farms and manors and two roads to nowhere. But it was a desirable place to 'weekend' or stay for pheasant and partridge shooting, being just a few miles from a rural train station with links to Reading and London. Many of the tiny, half-derelict cottages, often thatched and built of local brick or flint-seamed chalk cob 'clunch', had shallow foundations and no damp courses and were pulled down. Houses twice their height and size were built in their place. Four of them were our immediate neighbours. House prices and then rents around us rose inexorably, year on year. Our wages remained frozen.

Our bedroom looked out over the wood to the smooth green bascule of the downs: in winter, when the wind had blown holes through the wood, I could see Walbury Hill as well as the rise and soar of Gallows Down, also known as Inkpen Beacon or Combe Gibbet, all year round.

At the perfect height of Gallows' curve was a Neolithic long barrow, the burial site of the first farmers of the area. Built at least six thousand years ago, it was older than both Avebury and Stonehenge not far away. On top of the long barrow was a double hangman's gibbet, 25 feet high, cut from an oak grown in the village. A replica of a continuous line of six replacements for an original put up in 1676, the landmark is visible for many miles around. It was used just once. In March 1676, by order of Winchester Assizes, it was erected to display the bodies of two adulterous, murderous lovers, George Broomham and Dorothy Newman, 'in chaynes' near the place of the crime. George was a married farm labourer from Combe, the small hamlet at the foot of the south flank of the big hill, and Dorothy a widow from the village on its north (our) side. They met and conducted their illicit affair on the downs between their two villages. There are several versions of the story, but it is recorded that the couple had cause to do away with George's wife and son, Martha and Robert. Just beyond the long barrow, Martha and Robert Broomham were beaten to death 'each with a staffe' and their bodies thrown in the dew pond there.

Somehow the crime was uncovered and the two were brought to trial, found guilty and publically hanged at

Winchester Gaol. Their bodies were ordered to be returned to Inkpen to be laid out in the barn at the Crown and Garter Inn where they should be fitted with chains. A dispute arose over which parish was responsible for paying for, forging and erecting the iron chains and gibbet. The couple came from either side of the hill and the crime was committed on the chalk spine between the two parishes and (then) counties of Berkshire and Hampshire; but each of the borders, parish and county, stopped shy of the great ditches either side of the long barrow, making it a no-man's-island at the dizzying height of the hill. The parishes, who must have known the couple well, came together, split the cost, and shared the responsibility. George and Dorothy were taken back up the steep escarpment for the last time and hung in full view, at the highest point of the whole of Southern England.

The several accounts of the story include one written by author and naturalist W.H. Hudson (1841–1922) and another made into a film in 1948 by the young Oxford university student John Schlesinger, whose family lived in the village. *Black Legend* was the Oscar-winning director's first 'proper' film and he remained proud of it. Working with a very small budget, he involved local people from the start as well as his young friend, the actor Robert Hardy, who played 'mad Thomas', a character said to have witnessed the affair and the murders. Filmed in a fortnight at harvest time and just after the Second World War, many of the local actors continued their serious work in the fields between shoots. The fragile, silent, black-and-white film is extraordinarily atmospheric. It is shown every few years or so in the village hall and seems more poignant each time, as fewer cast members survive or others strain to recognise neighbours, family members or resemblances and notice how the fields and woods have changed. The gibbet on the mound on the hill, however, remains emblematic and, though its wood may have been renewed, the same.

Written into the tenancy of the most remote dwelling on the Estate, 17th-century Eastwick House, is a responsibility for maintaining the gibbet. In an unbroken tradition since 1676, strong local attachment ensured the structure was re-erected when it had blown down, been struck by lightning or rotted – but it has also been a landmark and potent apparatus for protest and political comment, up on high as it is. In 1965, it was sawn down in objection to the death penalty (by hanging), then was duly re-erected and sawn down again in 1969 for the same reasons. In 1970, Labour candidate Tim Sims hung an effigy 'in memory of the past injustice of a feudal system, brought to an end by a compassionate society and the never-ending fight by the common man for equality'. Appropriate, given that much of the downland around the gibbet had once been common land and agricultural Swing Riots, prompted by starvation wages, had swept the villages below 140 years earlier, resulting in burnt rickyards, broken machinery, threats to life and families destroyed. In the following decades, other effigies, political flags and posters were hung from the gibbet. In 1990, the plight of the agricultural labourer came to the fore again when an anonymous farmworker hung a scarecrow from the crossbar that dangled eerily by the neck. A board at the bottom of the structure explained how this 'agricultural worker' spent his £120 a week wage: £30 went on rent and £60 for food for his family, which left £30 for other necessities such as clothing, transport and other bills. 'How can I afford £14 per week for the Poll Tax? Is this the only way out?' His question hung, unanswered in the air.

Each time one of the seven gibbets was replaced, the names of those involved in cutting the oak, doing the carpentry work, digging again into the chalk barrow or erecting it recurred through the centuries. The current farmer's name of 'Carter' crops up again and again, as does 'Bulpit', a name that can be traced back since before the first gibbet was erected. In Newbury's museum, in the old cloth hall, there is a nearly 350-year-old 'chine' of bacon, roasted on the day George and Dorothy were displayed in their chains from the height to feed the crowd gathered there. Part of it was brought to the re-erection of the fourth gibbet in 1950 by the tenant of Eastwick. No doubt a Bulpit was present, then.

The gibbet still marks the place where lovers meet. It marks the place for summer picnics or Boxing-Day walks, for stargazing, kite flying, fireworks and birdwatching, for paragliding, horse riding and wild, windy welly walks, splashing through the great chalky pools and ruts that are like troughs of milk. It is a cyclists' challenge, a radio enthusiasts' high point, an occasional illegal rave venue; a viewpoint for a magnificent panorama. It marks the place where two long-distance paths begin – The Wayfarers and The Test Way – and, with its geographical collision of oceanic, ice-age chalk that faces North, South, East and West, it is a migratory highway for birds with a flora rare as a rainforest's. In storms, the gibbet soars like a ship's mast, the clouds whipping by like rent sails, the landscape below heaving, rising and falling, like the sea it once was. It is a landmark, a hotspot, a stopping place.

Mostly, I cannot imagine unquiet slumbers up there among the harebells and the bleating of sheep. Mostly it is a remote, unpeopled, sequestered place. Hard to find, hidden in plain sight – it draws the eye from miles around, yet people get lost and give up trying to find it.

From my bedroom, I have seen lightning strike the gibbet, meteors arc above it. The crisp black silhouette of the down against a clear sky after sunset masquerades as the ridge of the old threshing barn/schoolroom roof, as the poet Edward Thomas might have seen it, recorded in his poem 'The Barn and the Down', when 'the great down in the west' looked like a barn stored to its rafters 'with black of night.' Always atmospheric, at times the place has a wild, gathering and lonely air of tumult.

Next to our house was a cluster of very old cottages, angled with clear views to Gallows Down. One, uninhabited for decades, was grown over with brambles and volunteer ash trees, its tall chimney pots leaning away from each other like a V-sign to the newer, bigger, more comfortable and sanitary houses. Opposite was a 1950s house, contemporary with ours, with the pantry of an earlier house remaining as a workshop. Scratched and painted into its old brickwork was the date 1676. Both houses belonged to Mr Painting, another old, recurring name in the village. Several wild and unkempt cottages were owned and lived in by the lone men of the Painting family. Fiercely intelligent, guarded, sometimes eccentric or reclusive, they were respected and 'known' locally. Each morning, our elderly Mr Painting opened his curtains, stretched the sleeve of a jumper over the heel of his hand and rubbed a circle into the grime and condensation on the window to check the gibbet was still standing on the hill. Sometimes, he'd raise that hand in greeting.

In summer, I'd overhear him retelling old village stories to visiting family in the garden: tales of the violent uprisings of 1830 as vividly as if he was there – or with the still-fresh urgency of someone recounting it only secondhand. He also told his own version of the gibbet story: that the cuckolded wife Martha was murdered in his cottage, above the pantry gable that still stood. That the son had been alerted by his mother's bloodstain blooming on the ceiling (a stain that could never be painted out) and had met his end as he went up to investigate. That the date scratched into the centuries-softened brickwork marked a terrible justice served. This version of events is not corroborated anywhere else (though there are similarities to W.H. Hudson's story) and I wondered if he had confused it with Tess Durbeyfield and a hotel room in Sandbourne? Thomas Hardy's *Tess of the D'Urbervilles* killed her tormentor Alec and fled before her landlady discovered Alec's bloodstain spreading like a rose on her ceiling. Tess is captured at Stonehenge on nearby Salisbury Plain and hung at Winchester Gaol. Aged just 16, Thomas Hardy had watched the public hanging of Martha Brown at Winchester Gaol; a young woman who had killed her violent husband with an axe, grabbed as he whipped her. She went to the gallows composed, neat and dignified, and Hardy was haunted by the vision of her hanging as well as his subsequent shame in being there. Martha is said to be his inspiration for Tess. Yet, in some editions of his novels, a little hand-drawn gibbet is added to the frontispiece maps of Wessex, at Inkpen. Hardy was a regular visitor to 'Ingpen Beacon'. He climbed these lonely *Wessex Heights* to escape his own ghosts. 'Mind-chains', he wrote in a poem of the same name in 1896, 'do not clank when one's next neighbour is the sky.'

When old Mr Painting died, his house, its 17th-century pantry and the hidden, derelict cottage opposite were sold to a developer and pulled down. Just before, a procession of curios came out of the houses, including a Victorian Bath chair, a huge, ancient, ride-on stuffed dog on wheels and six ovens: each a vintage exemplar from a consecutive age. The developers took the roof off the house in spring, illegally destroying dozens of sparrow and house martin nests. The grass and mud cups landed with eggs and chicks among the brick dust, debris and shingle of roof tiles. I tried to shield the children from it all as we walked to school, but on the way home, contractors pulled up the ancient apple tree with a JCB as we were passing, partially toppling the old pantry. The huge globe of pink and white apple blossom exploded as it hit the ground and its scent flooded our nostrils. My son was outraged, but on this day, the developer made an eco-warrior of my inconsolable five-year-old daughter.

The hills were places to run to, satisfying a need and want for physical, mental and emotional freedom, a spontaneous and overwhelming desire to be wild, just on the edge of what might be socially acceptable. I was all in, dropping everything when it seemed to call me out.

Even when the children were small, I would take them up the hill, out late to see a spectacular sunset or a 'supermoon' rising, wobbling up from Reading 30 miles away like a bubble from a lava lamp. They have fallen asleep up there, cocooned in sleeping bags or rolled up in picnic blankets like human sausage rolls, among fragrant hummocks of wild thyme, calamint, marjoram and sheep poo, as we watched meteor showers or flocks of golden plover arrive, whistling in, in wild weather. On one occasion, when Rosie was six years old and she and I had a rare weekend to ourselves, we ate a picnic tea of cupcakes as lightning forked all around us, and we seemed safe in some kind of calm epicentre. A battered transit van drew up and a small circus spilled out – a juggler, a unicyclist, a guitarist and a hoop artist. We were their only delighted audience. We were often out late, irresponsibly, on school nights, after dark or squeezing a last hurrah out of the school holidays. They were the moments that mattered.

My dearest friend Sarah and I made wild, impetuous escapes up one hill or another, away from the hectic lives and demands of family home life. We met when our first babies were weeks old. We've spent evenings badger watching with the children at different stages – and sometimes not seen so much as a snail by moonlight. We'd go out at night with the gamekeeper in a Land Rover piled full of kids and straw and try and watch fox cubs – thwarted inevitably by the kids' hysterics. Though there were other once-in-a-lifetime sightings: a polecat hunting the down for rabbits, flowing in and out of holes like a bootlace through eyelets. Or a barn owl hunting right above a badger hunting for earthworms. We'd sit on the steep escarpment, hugging our knees, the whole valley spread like a cloth before us, watching the sun sink into wreathy mist or storms walk the valley like great grey gods, and drink tea from a flask or Pimms from cans like teenagers. And we'd buoy up broken, stressed, beloved teenage children as they negotiated the challenges, pressures and turmoils of young adulthood.

Best of all, we shared the ride of our across-the-field-neighbour's big chestnut mare, taking it in turns to ride Honey Bee up and over the hill, while the other pushed baby Rosie in the buggy. Nothing matched the wild, daft freedom and life-affirming joy of running, pell mell, pushing a pram with a bouncing, bonny baby in it behind a galloping horse ridden by your best friend. We would swap about halfway round when we found something to stand on to reach the stirrup of a 16.2-hand horse – a gate, a tree stump or staddlestone – breathless and laughing hard enough to test our still-recovering pelvic floors. When Rosie progressed to nursery, there was a bike – and, at different times, other horses to borrow. But red-gold

Honey Bee, her long flowing mane and tail, pricked ears and aloof, slightly amused, haughty manner was a constant. She and the hill and our various dogs.

Riding along the old drove road on the exposed white spine of hills and where the knobbly flints are like loose vertebrae, we were level with the kites, buzzards and ravens. It struck me that, hefted here, these birds owned nothing, but belonged absolutely. Here, on the great upsweep of thermals, the ravens perform breathtaking stunts and manoeuvres, seemingly for their own entertainment. They ply the updraft, hopping off a fence post in a synchronised manner in twos and threes and letting themselves be lifted up, before folding their wings and plummeting down to the ground, some thousand feet below, barrel-rolling as they go. They mirror each other in pairs so closely that, suddenly, one bird separates into two, a shadow that comes alive and peels off with its own free will. Sometimes, we are above them, looking down onto their backs, which is a rare view of such big birds. But eye level is best. The ravens in particular seem often to come alongside, eyeballing us in a quizzical, curious way, casually flip-flopping upside down, rowing a couple of beats backwards, the wrong way up.

―――

The neighbouring dome of Walbury Hill was, for a long time in history, famed as the South Country's only mountain at 1,011 feet above sea level (a 'mountain' in low-lying counties being considered over 1,000 feet). In the retriangulation of Britain in the late 1930s, Walbury was calculated at 974 feet. An almost-mountain. Still nine feet higher than Leith Hill in Surrey, only the Tors of Exmoor or Dartmoor and two heights (Hardy's Wills Neck and Black Down in the

Quantocks and Mendips, respectively) top it. The dew pond on Combe Hill, the collarbone to Walbury's shoulder, is reputed to be the highest water this side of Derbyshire. As far as chalk hills go, Walbury is the highest in the world.

There are times when I've sat, swinging my legs, on the trig point among wintering snipe and, on occasions, ring ouzel (a mountain blackbird to reappoint a mountain) or dotterel or snow bunting that were passing through, and looked down to the runway scar of Greenham Common. Or seen traffic flowing thickly on the bypass. I've stood on the long barrow and looked down towards my house, the clouds around my ankles. This intimate, sweeping landscape has its own microclimate of emotional weather.

In November 1830, those emotions overspilt, converged and ran like a river in spate, gathering force and volume as it went. Having had enough of poverty, hunger, job insecurity and arbitrariness, cap-doffing deference and damp, tumbledown cottages, the agricultural labourers rose up against their employer-landlords. William Cobbett was a household name. His *Political Register* would be read out in the local pubs to those who didn't have access to it or were illiterate. He visited the area regularly and conducted political meetings. One was held in the Pelican Inn, Newbury, in 1822; the very same inn in which the Speenhamland System (or Berkshire Bread Act) of poor relief was first devised and set in motion in 1795, to increasingly damaging effect. Two hundred people crowded the pub and, in a letter from Lord Carnarvon, MP of Highclere Castle (and Cobbett's old adversary), 'the doors and windows were besieged by the admirers of a man, who, whatever his faults may have been, deserved to be ranked as one of the boldest and purest of English politicians'. Cobbett's speech included the following:

> *The labourer has the first claim to the crop which the land produces for it is he that makes the crop – Crime does not apply itself to acts necessary to the preservation of life. God, nature, and the laws have said, that man shall not die of want in the midst of plenty...[the French Revolution] had not begun among the rabble, but among the quiet and dispersed labourers in the fields.*

A run of bad weather, exceptionally cold winters (heralded by intense, unfamiliar displays of the aurora borealis and the unprecedented arrival of huge flocks of wild geese), poor harvests and subsequent rises in the price of bread coincided with new, labour-saving inventions such as the threshing machine that would deny the poorest 'whop-straws' their winter income. Threshing was hard physical labour, but often the only work available in the lean, cold, hungry months. Many were unemployed from November to April and poor rates were not made available for short-term workers. The plots of land for common usage were relatively recently gone (as in John Clare's part of Northamptonshire). Wages were cut. Cottage rents increased. Anger grew at the profit-induced collaborative whim of the price of bread (for which they had grown corn) and the payment of increased tithes to the Church. Rural crime rates for poaching and food theft rose by 30 percent where recent enclosures of former common lands had taken place on the grandest scales, dispossessing labourers and their families of land use and living. Punishments for those caught were extreme and disproportionately harsh.

The machine-breaking, rick-torching riots began in Kent under the moniker of fictitious leader 'Captain Swing'; the name 'Swing' was derived from the action of the labourer's threshing arm and stick. With that work gone, the threat was perhaps implied that the stick be put to different use – which it was. In the villages below the big hill, as in other communities across Southern England and East Anglia, the labourers stirred and went from house to house, calling their neighbours out of the cottages until a force was raised against the landowners and the tenant-farmer masters. The crowd (including women and children) went from farm to farm, from big house to manor to vicarage, demanding fairness, humanity, bread, beer or money. Machines were ordered to be broken up, or were broken up by the crowd. From Inkpen, they went on to the small town of Hungerford where groups of angry labourers congregated and swelled to between 500 and 700 people. Some accounts put numbers at 1,000. Though no blood was spilt and no one injured, the riots here were among the most violent and damaging.

Some of the factions and groups were placated with a deal of advanced wages, while others were offered nothing; this was true for the cohort from Inkpen and its neighbouring village of Kintbury. The riots ran into days until they were eventually quelled by a rapidly recruited and largely self-appointed 'constabulary' of 300 men on horseback, led by Lord Craven. The day was described by landowner Colonel Dundas as 'a chase thro the country' and 'a good day's sport'. And although Kintbury's Reverend Fowle tried to prevent it, the men were literally hunted down. Particularly considering that only property was damaged, that no person was injured, the rioters and their families paid a very harsh price for their desperate actions.

Of the 18 local men tried, all but two were found guilty and sentenced to imprisonment or transportation. Three were given the death sentence. Of the 45 men sentenced to transportation in Berkshire, 24 were married with 78 children between them and several wives pregnant at the time. They were left to exist on the now-minimal parish relief of a 'pauper's dole', were separated from their children in the workhouse or assisted by the parish to move away. Nearly half of the transported men were from this portion of West Berkshire. In his book on the riots, *From Berkshire to Botany Bay* (undated, but around the early 1990s), Norman Fox, then Headteacher of the secondary school I now work in, stated the inevitable result of the harsh treatment of the rioters: 'The scythe of Whig justice…removed the hardiest and the best from the farming communities'. The villages below the big hill had lost many of their best farming men, their families plunged into greater poverty. Yet many of the employers, the farmers, magistrates, MPs and landed gentry were sympathetic to their cause, proclaiming most to be of good character and guilty only of demanding a humble wage and reduction in tithes and rent. Many were described during their trials as honest, trusty, hardworking and sober: kind and indulgent husbands and fathers. Lord Carnarvon stated in Parliament that the English labourer was 'reduced to a plight more abject than that of any race in Europe'. Petitions were raised and signed by 15,000 people across West Berkshire and managed to commute all the local death sentences but one to transportation. William Smith (alias Winterbourn) was considered the Kintbury 'Captain'. Reverend Fowle pleaded for Winterbourn's life and referenced his good character, but to no avail. Winterbourn was executed by hanging at Reading Gaol on the 11 January 1831, aged 33. He left a wife suffering with typhus and two children. After his death by hanging, Reverend Fowle took the unusual step of bringing Winterbourn's body back to Kintbury from Reading Gaol and had him buried in the churchyard with a large headstone.

I know I would have risen and gone with them, the rioters. My great-grandparents on both sides numbered among them itinerant agricultural labourers, shepherds and Romany gypsies. My Grandad certainly identified as a 'Romany Rai', a gypsy gentleman, with his raven-black oiled hair, leather waistcoat, paisley neckerchief and way with animals. He left me a smattering of Romani words, a fondness for fine china and woodsmoke, fresh air, poetry, stories, horses and hedgerows, a fear and respect of 'ghosties' and the conundrum of wanting home but feeling more at home out of doors.

My paternal grandmother's parents were known as Northamptonshire 'flitters'. Working and living on farms and moving on without notice, often at night, to escape debt or rent owed, or because Nan's father had offended landlord or master by standing up to them. When he sometimes lost his reputedly quick temper, they'd be forced to 'flit' in the night. When the mouths got too many to feed and Nan lived with her grandparents, she would often come home to find her family had moved on again. On one occasion, she arrived in the middle of a row over a stolen mangold at Orton in Northamptonshire. Her father, Harry Baxter, had been spotted lifting a stubble mangel-wurzel, or mangold, from the claggy, burnt umber soil to supplement a meagre pot of dinner on his way home. The farmer (his master) was told and Harry was confronted on the doorstep of Dropshort Cottage. In the ensuing row, Harry threw the mangold – a large, heavy root vegetable, not unlike a swede, and, as winter animal fodder, not even intended for human consumption – at his master's head, where it made considerable contact. Once again, he lost his job and the family home in one fell swoop, and with immediate notice. Though it is unclear whether he had already done so, before the mangel-wurzel was in the air.

Nan was a fierce, funny, feisty character, full of old country sayings and a protective armour of airs and graces, gleaned from an early life 'in service' to a big house. She also had a sharp tongue and distanced herself from the insecure life of an itinerant farm worker's daughter as soon as she could, saving hers and her husband's wages from the shoe factory where they both worked and surprising him one day by paying off the mortgage.

Landless, moved on, sometimes unhoused and hungry, certainly fenced out; I wondered if Nan's grandparents might have been among those rioters. Local parish records are full of punishments for trapping and stealing game on freshly enclosed land. The penalties were extraordinarily harsh and the criminals must have been desperate to risk everything to bolster a starvation diet. I have particular sympathy for them all, living, as I do, as a sometimes uneasy tenant on one of Cobbett and Clare's resented 'great game estates', a flint's lob from former common land. The actions to take from and punish the rural workforce are still, it can be argued, being felt today. How many generations does it take for those families to recover? To hope and strive for better? In the local secondary school where I work, between a third to a half of students' families are in receipt of benefits and free school meals. In the furthest, most rural part of the county, where there are also many private schools, second homes, big, country estates and houses, it is not a shock to discover among the third to a half, living in rented, tied or council-owned homes, family names shared with those Swing Rioters and 'ag labs'. Among some of them, aspiration and literacy levels are devastatingly low. Among others, tenacity, application and hard work win through, admirably.

I thought of those cottage dwellers, those desperate rioters when I heard ravens. Sometimes, an oracle bird sounded out a commentary and I was compelled to listen. Raven vocalisations are astonishing: guttural clucks and growls, repeated '*cloop, cloop, cloops*' like stones gone down a well, and a frog-like '*quark, quark*'. At times, the sound has a metallic quality, like an anvil ringing from some ghostly forge. Once one lone raven, away from his fellows who had found a dead sheep, called from a bank of larches within the park pale, his voice loud and confident. Was he calling out co-ordinates or summoning a coup or rebellion? He bounced from tree to tree, trying out different notes in different directions, wings folded behind his back, head down, as if he were a thoughtful speaker in the House of Commons for a parliament of ravens. I couldn't leave until I had heard all he had to say, this seeming-kingmaker that paced the boards, broadcasted the news. It seemed I must listen and try to understand. Even in the silences, there was the stiffened taffeta-rustle of wings.

The raven cocked his head at me as I drew nearer, and blinked his intelligent eye, and I wondered if this vocal experiment of his was a call-and-response exercise that I was failing? He ended his sermonising with a voice like a tolling bell that he adjusted to chime miraculously with the bell of St Michaels and All Angels in the village two miles away, just on the edge of my hearing.

I wondered, fancifully (gruesomely), if this speech maker, orator and sermoniser was descended from the birds that picked over the bones of the lovers that met on the hill. But in the intervening years, ravens were persecuted by farmers and gamekeepers and became extinct in the country, not long after those riots – except in remote, outlying colonies. A voice of dissent and incitement silenced for 150 years or so. But they are back now. These birds had likely recolonised from real mountains in Wales and have done well here, where there is such good living for scavengers. The number of pheasants killed on the road or elsewhere sustain an unnaturally high level of carrion-eating raptors and corvids, which are often the last straw for the remaining populations of ground-nesting, farmland birds.

I was once given a dead raven someone had found at the foot of the down, below the gibbet. Both wings were folded demurely across its breast and belly, crossing at the tips as if it were a dark angel, its head lowered on its breast in meek repose. Holding it in my two hands, I gently rolled it in the light and revealed so many colours other than black: magenta, emerald, royal blue and shining jet; magnificent, stately and transient colours of oil, preen and light.

The weight, solidity and size of the bird was breathtaking; it was as big as a buzzard. I pushed gently on the bird's wrist joint, as if it were a hidden button, to flex and fan the vast, black, fingered expanse of its wings. The mechanism moved easily. Then, I held the raven against me so that one wingtip brushed the ground: the other reached way past my waist.

Ravens can be long-lived (17-plus years in the wild), but they have their enemies. This bird's eyes, somewhat ironically, were gone. Chevrons of blood marked its chest like a coat of arms. One shoulder bone was cleanly broken and I could see inside the lightweight, hollow tube of bird bone and the supporting criss-cross of vaulted web.

It had been shot.

The bird's great, Roman-nosed bill, finely feathered around its nostrils and anvil-like, lent it nobility. I pressed my finger to the beak tip to feel the sharpness required to open carcasses.

Its skull and long, black, scythe-like wing feathers sit on my writing desk. A sort of Raven King for the imagination. Why is a raven like a writing desk, asked Lewis Carroll. When is a raven like a riot?

CHAPTER TWELVE

The Green Plover's Nest

Stories seem to want to repeat themselves, as we repeat them. Sometimes you have to not let them. The landscape itself seems to perpetuate them, to form their shape. The rest, perhaps, is up to us. Characters reappear with the same name, or the same job; connections are remade, broken and remade again. We are all made of stories. They roll round like thunder in a valley, or the seasons, just with less frequency and in altered forms, often not recognised for what they are until they are near their denouement. More erratic, they'll turn up like a horseshoe nail in the plough to puncture a tractor tyre or an ICI chemicals bag from the 1960s, kicked out of a badger sett more than a quarter of a century later. Nourished by the soil (and nourishing it), they connect and link like mycelia underground and pop up like fruiting bodies elsewhere, like the industrial, brown-rusted kerosene tanks peppering the dry downs, unearthed from Greenham Common during its restoration and now acting as water tanks for sheep and pheasants. Stories persist, like the re-erection of a gibbet through the centuries, or like wildflowers on overgrazed downland that spring into life when sheep grazing has been delayed by quarantine for blue tongue. One such April, the whole broad expanse of Gallows

Down was lit with golden cowslips packed so tightly you could not walk without treading on them. The glow could be seen from Oxfordshire.

The trick is to recognise when you're in the story and not be seduced by it.

Not to let the landscape take you where it thinks you should go. Repeating patterns, sending you down runnels like rainwater, shedding you off hills like sheep into combes, and tracks so worn and grooved they can be hard to get out of.

When I first came to the Newbury area as a child, living at Greenham and then Wash Common, Mum and Dad would bring us up to Walbury Hill or Gallows Down for walks. And when I got my first car, aged 21, I would drive up and spend whole days walking the tracks and ridgeway on my weekly day off from the stable yard. Compared to my memories of visits to the Combe Estate, back then, before we came to live here, its wildlife had certainly decreased. There were new woods, but they were understoried with laurel, snowberry and box honeysuckle: non-native plants used to 'warm up' a wood to hold pheasants in, but with little or no benefit to native wildlife.

The Estate's gamekeeper lived on the other side of the hill and told me he had a passion for wildlife and that he'd spent the last 18 years trying to bring it back. Our conversations as we bumped into one another in the farmyard, or down the lanes or on the footpaths, became more animated and grew into regular trips in the Land Rover around parts of the Estate I couldn't otherwise go. We went out 'lamping' at night with the million-candle lamp, a bright, handheld light primarily used for hunting after dark, but which the gamekeeper used, when I was there, to spot wildlife: foxes, badgers and polecats, or herds of fallow deer, the reflective tapetum discs in their eyes shining like strings of fairy lights in the dark. Sometimes the children would come, sometimes my friend Sarah, or the 'keeper's wife, my husband or another friend. But the 'keeper and I were more serious about it. We drove right alongside hunting barn owls, drew level with tawny owls and met our reflection in their black, dew-pond eyes, the lamp held askance so as not to dazzle them. We stood on the ramparts of the hillfort and watched short-eared owls sweep past, turning their kohl-rimmed, fiery eyes and cat faces to us as they went. We saw incredible, never-to-be-forgotten things, crept up on fox cubs playing, stalked woodcock at night, so that they were within touching distance, watched a flock of teal come in to one of the oldest dew ponds on wild nights and 'dreads' of golden plover sweep anxiously up in great gauzy flocks, like drift nets thrown into the sky. We had ringside seats at hare boxing matches, so close that when the raked fur flew, it passed right by our noses in wisps like thistledown.

During the autumn rut, we tracked a herd of wild fallow deer on the steep, wooded chalk scarp, where the soil was a moving scree of flint and chalk cobbles. We got close enough to see the thorny palmate antlers of enormous bucks, draped in elaborate headdresses of grass, stripped elder and baler twine; close enough to smell the testosterone-charged ammonia the animals had urinated out and rolled in. We heard their strange guttural calls through the wood and watched the nervy does and last year's fawns flee.

I still went out alone, perhaps increasingly so – or with the children – whenever I had the chance, but this wild world became close, closed and intimate. Something that the rest of the world often seemed outside of. We kept records of what we'd seen in a shared book and I wrote about much of it in my weekly newspaper column. I wanted to give something back to the place. To make my mark, making marks in a paper. I thought I could help bring species back with words. Conjure them up in the imagination of others who might give them a habitat, and make them real. The landscape became a muse.

On the hillfort in winter, at night, on the flat-top summit of our almost-mountain, we discovered birds that possibly no one else knew were there.

Past the dewpoint, trundling quietly along in the Land Rover, stopping, sweeping the lamp, we spotted dozens of feeding or roosting wading birds. There were woodcock, waddling over the open field like a meeting of Victorian philosophers, vicars or pontificators, cogitating, isolated in their own thoughts, long bills to the floor, wing-hands clasped behind their backs in deep thought, shirt-tails flipped up, gently bobbing with each step. Meticulously, slowly, they machine-sewed the ground with incredibly long bills, exploratory core-drilling; taking the earth's temperature. On some nights, we counted 40 woodcock on this field alone.

There were dozens of snipe, too, hunkered along the deep grooves of the old cart track. Nervier, they flew up on approach, with sweet, soft, rasping calls in the night. They were answered by the plaintive, fearful piping of golden plover. None of them went far, rising and falling back down to the earth. I slipped down from the Land Rover and crept up on a woodcock on the very edge of the pool of lamplight, so as not to dazzle it. Its large, intelligent eye – set high and far back on its head – glinted like a tiny onyx in the night. A flick of the lamp away and it was hard to tell what was reflected: a glint-full field of wader's eyes or a lamplight's bounce back from the dew fall. A field full of tiny stars. Crawling on my knees, I got about as close as it's possible to get to a woodcock in the grass. It sat tight like a wooden carving, right next to my hand. I could see astonishing detail in the cryptic, owly, dead-leaf-and-grass feathers. I could see the sensitive, mobile bump at the end of its long bill. I saw it move, almost imperceptibly. Tears streamed down my face in the cold wind, where I dared not take my eyes off the bird. Eventually, I had to stand up. When I pushed off the ground with the flat palm of my hand, water oozed through the stitch-holes made by the birds' bills, between my splayed fingers. The woodcock got up and waddled forward, a clod of earth and mulch come alive, a jigsaw piece of grass that got up to walk beside me. A creature absolutely inseparable from its environment. It wasn't until I had to step back up into the Land Rover in my soaking jeans that it flew up vertically, hovering and illuminated like an angel above me in the lamp's beam, before landing out of sight.

We went quietly along until we saw the fourth species of wader we'd hoped was still here. An unlikely bird, and a rare one at that, it took us both by surprise when we saw it first, seemingly far from its own habitat. It took each of us time leafing through books and searching online until we were both, independently, certain. But here they were, in plain sight among far more likely birds. Ruff.

It was an exciting, thrilling moment, this confirmation.

In winter, ruff are unremarkable, long-legged, brown waders. But in summer the breeding male develops a spectacular Elizabethan ruff and ear tufts in such variable piebald and brown colours that no two are alike. They dance like lace-cuffed dandies at each other to win a mate. But this was winter. Stockier than snipe, ruff have long necks, small heads and a shorter, faintly downcurved bill. They move somewhat apologetically – the muntjac of the bird world.

These ostensibly similar birds tested our ID skills, and I was always a beat behind my companion until I got my eye in. Snipe flew fast, zigzagging and calling, white undercarriage to the frosting ground, while bulkier woodcock went straight up, just once, and landed close by. Ruff were an entirely new species for us both, complicating and thrilling the night. Their flight was deeper, slower, lower; more hesitant.

For several consecutive nights we went up to see them and also saw short-eared owls and barn owls. It felt both wildly exciting and poignant. There was an air of 'last chance to see' about the birds and in discovering them came a particular sense of responsibility.

We learnt of plans to mow, tidy and intensively graze the hilltop. To fell surrounding trees, gorse and scrub. There were several agencies at play. The shepherd needed better grazing and hated the gorse that tangled and trapped the sheep. Natural England wanted to increase the open chalk downland and English Heritage wanted to preserve the hand-dug ramparts from tree-root damage. We threw our hats into the ring, too; how would all this affect the birds that seemingly no one but us knew were there? We tried attracting the attention of the county Bird Recorder, but he dismissed our sightings as highly unlikely. Proof in the way of photographs proved beyond us. On the nights I wasn't up there, I couldn't sleep.

Under a new young farm manager, the Estate began to create more habitat for wildlife, through the agricultural Higher Level Stewardship Scheme. On 'difficult' ground, corners and headlands, nectar strips were planted for pollinating insects and seed-bearing cover crops just for the wild birds. Around awkward lines of telegraph poles or just to form a break across huge exposed and hostile acres, beetle banks were included – heaped, weedy furrows in narrow strips that created sheltered corridors for invertebrates and all manner of wildlife. Skylark plots – small squares where the seed drill was lifted for a matter of seconds in autumn-sown crops – were left in fields, leaving nesting and foraging areas for these quintessential farmland birds. In other areas, stone curlew plots were created: areas of land, halved, with each side alternately rough-ploughed or left, early on each year. This left a weedy, open habitat for these rare birds and also proved a sanctuary for lapwing, skylark, finches and buntings, butterflies, moths, brown hare, voles and glow-worms.

Most excitingly, a small seed hopper was bought to fit the back of a four-wheel-drive Gator, a small, all-terrain vehicle, and it was used to spin out birdseed onto the cover plots from late autumn through to March. This covered the 'hungry gap', when many songbirds in the modern countryside starve. It was a favourite job to plunge my arms up to their elbows in the cool gold treasure of linseed, rapeseed, millet and sunflower: to go out in the roaring machine, flick the switch and watch the gold dust spin out behind me and the birds descend to feed.

I became a sort of unofficial, voluntary wildlife advocate on the Combe Estate. I tried to influence what I could, suggesting different ways of doing things in addition to what the farm was paid to do under the Stewardship Scheme. I pointed out the damage caused by unnecessary actions undertaken at the wrong time of year and the positives (and economic benefits) of doing them less, or not at all: cutting hedges once every three years, or on the road side only, rather than several times a year. Not mowing remote field margins and headlands no one used. Flashes and shallow ponds were dug, new hedges and woods planned and planted, and

I felt part of the decision-making process in what should go where. I begged, pleaded, wrote, nagged, motivated, entreated, implored – and challenged the gamekeeper, the two farmers, the estate manager and the landowner. Often, I toed a fine line. I tried hard to be upbeat, positive and jolly; beguiling all I could with the sensuous wonder of the wildlife *they could have*, and the economic sense of time and money saved. I tried hard not to show my frustration and despair when they continued how they always had done. I persisted with a quiet, determined voice, knowing all the time that these people worked for those, or *were* those, that owned our house. We couldn't afford for me to make a nuisance of myself.

With every success came a series of exasperating failures. Beetle banks that became thick, glorious masses of vegetation full of life (including small coveys of grey partridge) were forgotten about when the field was sprayed with Roundup weedkiller and died. New hedges were put in, while old ones full of ancient, twisted hawthorn were killed off by drift from the spraying arm or butchered over and over by the flail, several times a year, so they no longer had a chance to bud, blossom and fruit. Time and again, we would have to start again. All too regularly, an overfilled spraying tank would be run out over 'the grass', killing everything.

The nectar strips became tall plots of brilliant colour, loud with bees and bright with clouds of butterflies, next to vast acres of factory-floor-perfect wheat without any life in them. Great swathes of downland were over- or under-grazed, or simply grazed at the wrong time of year, so the profusion of rare chalk grassland plants were patchy or eaten off, leaving nothing for the insects that relied on them. The crops were sprayed endlessly with chemicals. Industrial quantities of metaldehyde slug pellets were spun out from the back of a tractor over the fine, pale, apple-crumble tilth of the chalk fields and their new skylark plots, giving them a lavender haze of poison. Chemically pre-coated cereal seeds sat like bright orange and pink sweets on the surface; all accessible to hungry birds. Each autumn, a pre-emergent spray stained the flints yellow, the smell lingering for days, before any shoots had even breached the soil. The hares licked their feet and died. Like the farms all around, it was a toxic landscape.

A small, triangular piece of open wood near our house was one of the last strongholds for adder, hedgehog and silver-washed fritillary butterflies in the village. One year, and for several after, it was included in a June mow, just when the wood was at its most tender and fecund and held the most life in it; just when the hedgehogs had their young and the butterflies were emerging into the dappled light and the wildflowers were at their best. It was mown, not without difficulty, to the ground; hibernaculum stumps smashed and exposed, the spinning mower blades slicing it down to bare soil. The arisings, or cuttings, were left to enrich the soil and encourage coarser plants like nettles and docks to grow in place of the wildflowers. Despite my repeated handwringing, the message didn't seem to get across. The populations of those animals dwindled to nothing.

I tried to concentrate on the extraordinary success of unsprayed beetle banks and the great swathes of wildflower and nectar strips. And they were incredible. They worked. They really, really worked.

One particular acreage of bird cover was planted up with linseed, quinoa, kale, millet and mustard along with triticale, a wheat-rye hybrid that bore a heavy seed, but also held the whole crop up off the ground with its strong, upright stalks, so nothing rotted. It also ripened consecutively, providing seed in an otherwise impoverished countryside. One snowy day, the 'keeper and I went up to see it in the Land Rover and, as we rounded the corner, it exploded into breathtaking life: the numbers of tits, finches and buntings thick as starlings above a reed bed. Every square metre moved with feeding, twittering birds, hanging upside down on bent seedheads that catapulted snow when the bird left.

Sometimes, a sparrowhawk or peregrine spooked the birds and they erupted in a bouncing, twinkling mass to re-leaf bare trees on the wood edge, before swooping back in a continuous stream to feed again, tawny bodies warm against the white snowfield. The sound of so many pairs of wings and their chiming voices was that of a great aviary.

Among the constant movement, we tried to pick out individuals and discovered huge numbers of chaffinch, greenfinch, siskin, yellowhammer and linnet, but also lesser redpoll, reed bunting, brambling and, possibly, tree sparrow.

This plot and others like it had been long in the planning and dreaming. These were flocks of the size and appearance my grandad told me about. This was nothing short of the realisation of a dream and the worth of a field of flowers to a thousand little birds.

The management of the Combe Estate as a whole was no different to any other of the big shooting and arable estates adjoining it, and then each other, for rolling, green miles into Hampshire and Wiltshire. This was normal, accepted, farming practice. Some (like us) went in for Higher (or Entry) Level Stewardship schemes, where they were paid to provide or create habitats for wildlife, but around and outside of those areas, the damaging cycle of intensive farming and the management of estates for shooting and forestry continued, often to the detriment of the wildlife, which plummeted. Species were extinguished. Woods were clear-felled during the bird-nesting season, hedges routinely and harshly cut (even in spring), and verges and field margins cut in summer, needlessly. Even on our farm, the cutting, mowing, weedkilling and tidying went on all year round, whether anyone used or even saw those areas of the Estate or not. Some of the neighbouring landowners owned several estates – even whole villages – and did not live there. On this family-owned estate, which was loved and lived on, there seemed a genuine interest, but no one to take the lead and push its special wildlife to the fore and include it in every decision. The wildlife simply wasn't *known*. I wondered, could someone who doesn't own or manage the land really have an influence? Someone whose only claim came through love and an intimate knowledge and understanding of the place?

For a while, the 'keeper and I kept a shared record in a little green, soft-covered notebook of the bird species we'd seen – and passed it on annually to the local Bird Recorder. The book would travel back and forth over the hill and we'd message each other sightings to write in. Apart from occasional records from birders, we seemed to be the only ones noticing and recording the wildlife here, regularly. By writing about it – in my local newspaper column and later by blogging – I was the only one publishing the changes, the losses, the few successes. By bearing witness to the wildlife, there was an opportunity to resist its loss, call for action and responsibility, and influence change. In that way, the traditional roles were reversed: the landowner became just another tenant, required to take a different kind of ownership. A responsibility for managing the wildlife *known* to be there, and that belonged to everyone and no one.

———

In winter, lapwing flocks of around 80 birds could be seen over 'Ninety Acres' Field, and more on Sugglestone Down. But in spring, they were reduced to a remnant breeding flock of around 20, at a few different locations. Their numbers were critical. In Southern England, the lapwing, or green plover as the gamekeeper called it, is quintessentially a farmland bird. Elsewhere in Britain, it is also a bird of estuaries and wetland, open moor and heathland, with its own challenges. But its Southern English homeland had become hostile to it. The lapwing and its nest are fully protected by law. The only exception is legitimate farming practice. Lapwing decline has been greatest in Southern England where farmland is their only suitable habitat. In just ten years (1987–1997) numbers dropped by 49 percent and have been freefalling ever since.

They are my favourite bird; a bird that has evolved with the worked land and the men and women who worked alongside it to such an extent that it has the most surviving vernacular 'country names' of any other bird: peewit, pyewipe, horneywink, flopwing and teewhuppo, to name just a few. 'Lapwing' is thought to come from an old English word *hleapewince*, meaning a leap with a wink in it, which describes their flight perfectly. The names describe variously the colours, flight or call of the bird, which all vie for attention. I switched easily between green plover, peewit and lapwing, readily using all three without thinking, depending on what glancing light the bird was seen in or whether it was calling, or in flight.

The Green Plover's Nest

Green plover have a refined elegance about them, but equally, something of the farmworker's robustness. Perched on a chunk of turned earth, they are upright, alert, smartly smocked in black and white, with a long, fine crest. But in sunlight, they gleam petrol-green and teal, with a wink of magenta.

They wheel and dive on the strong black-and-white aprons of wings whose tips are wider than their base, tumbling with a walloping snap and a warbling yodel. The soaring '*pewit, wit, wit-eeze wit*' that follows is quite the loveliest, most joyful sound. It is a bubbling, looping, excitable cry; distinct, melodic and like nothing else. Interspersed with a sound like a stick played down a comb – or a wooden, ridged guiro – it is the aural interpretation of a stomach-flip of excitement. To my son Billy, they sounded like a whale or an electronic yo-yo. Close to, and you can hear the creaking beat of their owl-like wings.

Being a ground-nesting bird of open country, the hen lays four eggs scattered within half a metre. When she is ready to incubate, she rolls them into the nest, little more than a shallow-scraped saucer, shaped by the breast of the male and lined with a few strands of grass. The hen bird turns the points of the eggs inwards, making their circumference within the nest as small as possible. Coloured and shaped like pieces of chalk-nibbed earth and flint, the eggs are almost impossible to spot. Years ago, when these birds were valued more highly, the nests were marked and horse, then tractor, was halted, the nest moved a furrow over and then replaced once the machinery had passed. Or it was avoided. Often, a single egg was pocketed for eating, from each clutch. In an attempt to protect the eggs and chicks, the gamekeeper and I took to walking the furrows of the arable fields in spring, to locate and mark the nests with sticks and small flags, alerting the two tractor drivers. Neither were aware they were driving over nests or even that these birds nested on the ground.

But the flags did not work. Clever corvids worked out that the little flagged sticks marked an easy meal and some of the nests were predated. We moved the sticks further off. We worried about other predators. The plot was criss-crossed with badger trails and there were fox cub droppings between the nests. We planned an electric fence deterrent. Lapwing rely on the early warning system afforded by a pair of eyes with 360-degree vision. If a predator approaches, one of each pair flies up to mob the target, on the ground or in the air, twisting, shouting and beating those broad wings. But it takes a flock of birds to do this, a critical mass to put a predator off. Four, five or even six birds just won't cut it.

We had some success on Sugglestone Down and chicks hatched. They were up and running straightaway; black-and-olive spotted pom-poms on strong, long, precocious wader legs. They froze instinctively as the shadow of a bird passed over and became invisible in the furrows. A joint team from the RSPB and the Game and Wildlife Conservation Trust came to catch, weigh, ring and release the chicks and I got, briefly, to hold them. But they were so vulnerable. The sky was full of crows, rooks, ravens and jackdaws, buzzards and kites, the predator-prey ratio kept artificially high with the thousands of pheasant and red-legged partridge bred and released for the shoot each spring and summer.

One very early morning, we went up to the stone curlew plot where they had been nesting and, through binoculars, took in the sweep of a long crest and a bottle-green back in the sun. There were nine birds. But before we began to revel in this spectacle, it became obvious to both of us at the same time that something was very wrong. The birds were restless.

The Green Plover's Nest

Agitated. The docks and weeds essential to this sanctuary were curled inwards at the edge and there were fresh tramlines over the entire area.

With a growing sense of horror, it dawned on us that the plot had been sprayed with herbicide; whether through accident or misunderstanding, we never did find out, but it was clear that the tractor and its wide spraying arms had gone right over the lot.

I returned to the field the following evening, alone. The sunset was glorious. The electric fence to ward off ground predators would not be needed now. The raven calling from the wood below a fingernail moon posed no threat. The damage had been done; the lapwings would not make another nesting attempt that year. I resisted checking for sure: for eggs, cold between furrows, or yolks soaked into the ground. Chicks. The silence is loud enough. The haunting, fluting, far-off cry of a lapwing is all in my head.

CHAPTER THIRTEEN

Rural Work:
The Dew Pond on the Height

Some stories are reabsorbed into the landscape and gone like the lost 'white horse' of neighbouring Ham Hill, the thin green turf growing back over the chalk scars of a curved neck and withers, the angle of hock and fetlock. Those members of the Bloomsbury Set, hiding out in this small, remote downland village, might still have known it. They trekked up the steep, orchidy slope to picnic there, scandalising (possibly) the locals with their naked expression, bisexuality and the complicated geometry of their relationships. (The locals may have just been grateful for the work, the opportunity for entertainment and a validation from the 'upper classes'.) But stories do and have left their mark on the narrative history of the landscape; a stain upon it, like the strip lynchets of an old farming system that resurfaced in hot, dry spells or old, wall-sized maps discovered hidden in the cellar at the Big House on the Combe Estate, marked with great patches of common land, pre-enclosure, or on brand new maps, bought to replace ones worn soft as cloth, which revealed new open-access land, glowing at the edges of previously well-kept, fenced-off secrets.

Rural Work: The Dew Pond on the Height

Snowdrops and yew trees concealed the footprint of an old cottage and a flint-lined well, and bomb craters, indistinguishable from dry dew ponds, or Bronze Age 'pond' barrows shelter deer, hares and flocks of golden plover. There were the burn scars left by stolen, torched cars and gouge marks in the ramparts where the gamekeeper's Land Rover was pushed off the hill after a monumental rave.

Below the hill in summer, thin green lines run across the stubble, like green crayon on yellow paper, where the corn has been continually trampled by the night traffic of badgers and the tougher grass has grown through. Even when the field is ploughed, the trail reappears, trotted out like a watermark returning; an ancestral, indelible line. You could track it for a mile as it runs up and over the rounded hill, where it turns pale; a seam stitched over the hill's breast, a silver stretchmark, a badger's desire path. The white chalk horse has never reappeared, though.

The feeling of being a small part of this great landscape's story, and the constant highs and lows of its conservation rollercoaster, kept me engaged and on a taut, singing wire. The juxtaposition of gain and loss, the lurch from despair and frustration to exhilaration and delight, led me to believe that if I just kept going, kept trying, I could change things, despite repeated setbacks. I was hooked. I couldn't let go. My love for this landscape has rewarded and hurt in equal measure. I was all in.

We managed to hang on to some of the scrub and gorse on the hillfort and limit the sheep grazing. A compromise was struck that suited all parties. The following year the birds returned: snipe, woodcock, golden plover, short-eared owl, barn owl and even the ruff and a small number of grey partridge. But the more I looked into things, the more I could see that here, and on almost all other farms around, an inexorable mass extinction of agricultural wildlife – wildlife that had evolved to live alongside us and that had been celebrated throughout our culture and literature – was well underway. An ecological holocaust was taking place. The contract farmer alone, with his knapsack weedkiller or the sprayer, with mower or hedgecutter, was totally indiscriminate, mowing at the most sensitive times, cutting hedges after dark, when birds had gone to roost, or spraying on windy days – a one-man environmental disaster area. The estate manager, who had had his eye on things and who had introduced widespread conservation measures, was posted abroad. Suddenly, everything was sacred, nothing was safe.

In everything I did, I tried to spread the word. I took writing and wildlife workshops into schools (and out of them) and the 'keeper and I did talks and guided walks, attempting to show how a farming and shooting estate could support wildlife too. We did a series of talks for some influential American ladies who were guests at the Big House. It was a hopeful attempt, and one I believed in. There was plenty of push and pull during our talks, plenty of banter, open disagreement and good-humoured challenge. I thought, ultimately, it was working.

Yet all the while there hovered the precariousness of a tenanted life. The tentative influence I had on the country estate relied on toeing the line, on being non-confrontational, on the largely feminine wiles of thoughtful gentleness and suggestion, patience, flattery and positivity. Sometimes it galled. I could not deny that this was, first and foremost, a farming and shooting estate, and I had to acknowledge that. I could choose to walk through it and the community I was part of, railing against wildlife-denuding practices, or I could

Rural Work: The Dew Pond on the Height

engage with it, get people onside and persuade them to find a better way to allow, include and encourage more wildlife. I even went beating – the part of a day's shoot where a team of 'beaters' walks, pushes and eventually flushes pheasants in from whole hillsides away, so that they fly over a line of guns and are shot at. I had a complicated and difficult relationship with it. On the one hand, the day involved walking through the heart of challenging and beautiful terrain in all weathers with my dog, spotting wildlife and sharing the thrill with a rural, working and knowledgeable community, while on the other hand, there were the nonsensical ethics of shooting on an increasingly commercial scale – for 'sport'.

I could convince myself that it was done well here; each 'gun' was given a brace or more of oven-ready birds from the previous shoot, while the rest were sent to restaurants or the game dealer. It helped knowing that the extensive acreage of wild bird cover and nectar strips, the new copses and hedges planted, the beetle banks and field margins were all in place because of the shoot. That the purchase of additional tons of wild birdseed, the seed hopper and the gamekeeper's time to feed the songbirds from October to March was funded by the shoot; that scrapes were dug in the woods or on the edge of fields to provide a habitat for waders (not to be shot). That I would never shoot anything. That the pheasants were caught up and bred from here each spring; that ravens, kites and buzzards nested unmolested in trees inside some of the release pens and were welcomed – by then, the poults were mostly too big to be taken, and if some were, what of it? The fact that the presence of goshawks and sparrowhawks was celebrated. The fact that those animals traditionally considered 'vermin', such as polecats, stoats and weasels, magpies and jays, and even those foxes that weren't an immediate 'problem', were not only tolerated, but their presence accepted and enjoyed. Apart from the pheasants and partridge bred on the estate, no other 'game birds' were shot.

And the money came in handy. It paid for Christmas. It meant there was a little teapot stuffed with notes to fall back on and our winter weekly shop was stretched out with many a pheasant casserole. Often, it was the only meat we had.

It was work. I found wry comfort in discovering that John Clare, the agricultural labouring poet, worked with the enclosure gangs – the very enclosures he lamented. Enclosure brought an opportunity of employment Clare could not turn down: fencing, hedging, the destruction of trees or lime-burning – all to enclose the former freedoms of his own parish. It also brought fresh faces and new opportunities to socialise. Clare scholar Professor Simon Kövesi wrote in his book *John Clare: Nature, Criticism, History*: 'There was no economic space for Clare to consider *not* doing the paid work of enclosing his village, or of lime-burning; choice is a product of socio-economic power, and he had none. There was no front for resistance because poverty denied space for that activity'.

We were not living in poverty. We hovered just above the line where benefits kicked in. But things were undeniably tight and stressful at regular intervals. There was no give, no buffer. Like John Clare, we didn't really own anything. We had bookshelves of books, some clothes, photographs and bits of furniture; minimal crockery, four pans, three wine glasses and enough cutlery for seven. The curtains, carpets and oven weren't even ours. My parents lent us money to buy a car and bought us winter tanks of heating oil and mattresses for the children's beds; my parents-in-law bought the children's school shoes. Neither my husband nor I had life insurance or passports. Even the horses we rode and looked after every

Rural Work: The Dew Pond on the Height

day were borrowed and I worked part-time in a library – whose whole ethos was borrowing. I cleaned for one of the big houses, owned by an extremely wealthy family who used the house for occasional weekends and breaks. I was required to keep it top-end hotel standard and all worked well for a while, but the pay was erratic and I could never quite shake off a feeling that I was being watched. Written notes on how to do things a particular way (and not the way I had been doing them) were particularly unsettling, as were the times when the chauffeur-butler suddenly materialised when the family were away, at one of their other homes. On one occasion, when I hadn't been paid for three months, I picked up over £300 worth of notes strewn over the bedroom carpet. Another time, I collected up loose pages from an event the family were organising that appeared to contain the mobile numbers of celebrities and the nannies of royal children. I carefully stacked the cash and the pages on the bedside cabinet and waited another few weeks for my money. A lot of good food got thrown away (until I made the decision to take it home) and I'd wash, iron and hang up beautiful clothes that my employer's children would have grown out of the next time they visited.

My home was not mine. With a change of someone else's heart or plan, or a rent hike, we could effectively be homeless with three children and a dog; a whole community built up and belonged to – gone in a couple of months' notice. There was nothing else we could afford to rent locally. It sometimes felt like a precarious existence, and had been so since I left home, almost 30 years before.

Kövesi mentions Clare's enjoyment of the company he kept with the enclosure gangs; the socialising and the drinking. However, along with the effect of bone-tiring work, he lamented the distraction from his writing. We enjoyed the camaraderie of the beaters' company too: it was very separate from the paying guns – we barely came into contact with them and rarely spoke. If our paths did cross, there might be a nod, a 'good morning' or an occasional thank you. It was routine to be ignored. But I enjoyed the feeling of belonging to a motley crew of raggedly dressed, deeply rural people of all ages; the (sometimes tongue-in-cheek) country lore and connections to the land that went back generations. I loved the comfort of the pub afterwards, or the fire; wet dogs and coats steaming, discussing, mostly, the wildlife we'd seen. But I grew increasingly uncomfortable with it; the bits that didn't add up. In particular, the unacknowledged, unregulated and increasing scale of it as an 'industry' and its unquantified impact on native flora and fauna. At dusk, when the pheasants were clumsily going to roost in the trees, the cacophony of cock birds coughing out their alarms triggered all the others in the vicinity, from estate to estate, into Hampshire, Wiltshire and on. The sound echoing and reverberating off the hills and rolling round the combes was an overwhelming wall, drowning out all other sound, louder than the big tank guns that thundered away on Salisbury Plain – and set the birds off again.

Meanwhile, the pull of the hill grew stronger. I didn't always know how to manage it. Up there, I was wild and free, absorbed in something else: something that was not domestic routine and always being behind and not *house*. It was an escape, a ducking out, a vanishing, a beyond-the-ticking-of-the-clock. It was reckless and it was not being called – or, it was being called by something else. It was a kind of work.

The gamekeeper often turned up, showing me things that he had spotted, working out there all hours: a long-eared owl, a

Rural Work: The Dew Pond on the Height

big flock of wintering lapwing, a Montagu's harrier, a white roe doe, a collapsed badger sett that revealed all its inner workings like a cross-section diagram, stone curlew chicks. And I spotted plenty of things too: sometimes keeping them to myself, sometimes triggering an excited volley of text messages.

It was wildly exciting, thrilling and seductive, being out there. Addictive even. From my bedroom window, restless and up late when Martin was on another set of punishing twelve- or thirteen-hour night shifts, I'd watch the gamekeeper's lamp raking the sky, sweeping the barn-roof silhouette of the hill. Sometimes its uninvited beam would wake me, sliding into the room between the gap in the curtains as if it were the moon, falling onto the duvet cover, glancing off the mirror, and I'd squeeze my eyes shut, willing it to go away. I couldn't go out – couldn't leave the children, of course – and I wondered what I was missing up there: the lamp stalling on a string of fairy lights that materialised into the reflection of the paired eyes of a herd of a dozen fallow deer or a pair of barn owls hunting at close quarters. Up there, I felt part of something ancient and was unafraid; there was the handful of scattered lights from each village below, either side of me, and the town's lights some nine miles distant – otherwise, such deep, velvet darkness.

I felt safe up there, knowing that the real light was down in the valley and not up on the hill at all; real, steadfast and glowing – the reading light in the girls' shared bedroom, my son's torch; later, the light from their phones, the fire lit, the blue TV screen flicker of Martin watching football. Never faltering, even when they'd long been blown out, put out, turned off, by the time I'd got home.

At times, I was still too gripped by that strange outsideness to go in. Even when I got back to the car-park corner of our home field, I felt too wild; almost cruel, selfish. I wouldn't go in; I wouldn't go home. Not just yet, anyway.

The chalky smudge of the Milky Way unfurled above a field lit faintly by starlight. There was Castor and Pollux, Cassiopeia; the bright red bull's-eye of Taurus's Aldebaran and the hot coal of Betelgeuse in the hand of Orion, as he cartwheeled over the hill. Sometimes, above the down, a fingernail moon hung like the curved bevelled edge of a clockface, Venus swinging below: hornlight, hoarlight.

I think I was only testing. Testing solid ground. Pushing against a five-bar field gate because I felt safe enough to do so, confined as I was to living week after week within a five-mile radius. A bit like leaning over a parapet, wondering what it might be like – how terrible it would be – to fall. It was a kind of security, perhaps. But when the wind rose, or I came back from doing the horses at dusk, the feeling, the calling would be there again. The blackbird's shrill and persistent '*pink, pink, pink*' was a voice like knapped flints, chipped off like flakes of flint into my heart. I expected blue sparks in the dark.

One winter's night, I played the part of an injured woman lost on the big down at night. Martin had joined 'HART', the NHS specialist paramedic Hazardous Area Response Team. This meant he was trained to work in the 'hot zone' of dangerous incidents which might involve fire, firearms, hazardous material and chemicals, water rescue, heights, collapsed buildings – or steep, remote hillsides at night. Regular live training scenarios were a big part of the job and, as our field car park filled up with rescue vehicles and people, I headed up into the snow to lose myself on Rivars Down. Rivars (pronounced with a broad Berkshire *Ry vars*) flows on from Gallows Down and is mostly wooded. It is

Rural Work: The Dew Pond on the Height

very steep and the chalk soil so thin that it is not much more than the previous year's leaf mulch on top of a scree of loose chalk and flint. It is almost impossible to walk its length without sliding to your knees at least once. Any desperately grabbed handhold is a risky business – elder is likely to come out of the ground, the hazel that grows is unusually brittle, and you risk a sharp and penetrating response from hawthorn and a nasty infection from the blackthorn. Apart from the bare, green picnic rug of a gap in the middle, there is no public access and no phone signal. It was the perfect place to pretend to be a difficult-to-access patient. I went halfway down, lay on my coat in the snow, wedged my feet against a hornbeam stump and waited. Tawny owls floated through the branches of the starlit wood, calling. Below me, a shadowy herd of fallow deer moved through the trees, backlit with snow. I saw more and more as my eyes became accustomed to the dark. The humpy form of a badger walked along his own path, ten metres down the slope, his silvery fur and black-and-white stripes suddenly making sense as a cryptic form of camouflage. I heard a vehicle go by on the ridgeway above and, knowing Martin had set his teammates off at the other end of the hill, I settled in for a long wait. I heard a deer banging the corrugated tin lid of a pheasant feed hopper as it licked up the spilt grain and watched rats emerge from holes under the roots of a beech tree. I began to hear people calling from far away and it took me a while before I realised it was me they were calling for – and then another while before I thought I ought to answer. But at that moment, a fox sauntered out from the shadows and appeared right in front of me. It sniffed and spooked at my boots, then took a couple of steps back towards me. We held each other's gaze for a moment in the monochrome wood, until the fox went on its way, unperturbed. I felt elated. I felt hidden away and comfortable among this otherworldliness – lost to my world and accepted by this one; unfound. The calls came closer and still I remained mute. It wasn't until they drifted further and further away that I managed to gather myself together, made a real effort and called out, letting myself be found, put into a splint-type stretcher and hauled up the hill by the six-wheeled Polaris all-terrain vehicle. In more ways than one, it was a realistic scenario.

On the hill in the snow, I'd begun to experience a paradox; a growing feeling that my universe was shrinking, but at the same time expanding to fill the dark night skies, the immense hanging woods and all the landscape below. The calcareous grassland, its soil and turf so thin, what lay beneath was barely concealed: a whole, heaving ocean, calcified. Surely, everything I needed was here. I had become environmentally, acutely, sensitive. But at the same time, might I have stayed out too long? Gone too native? Had this landscape become too like a person to me? If it had, I was part of it; porous chalk with seams of flint, a chip of ice at my heart, a sea inside.

And then I'd think, of course not. I'm a fully present and best-I-can-be wife and mother. This is just what I do when I'm not that.

Out alone, I began avoiding the 'keeper, wanting solitude when I could get out – and found that this wasn't easy. I started to behave more like a prey animal. Walking at dusk was filled with a frisson of alertness and fear and, in the half-light, a dead, hedgerow elm wrapped in an unravelling knit of ivy looked for all the world like a hunter with a gun or an old god – Herne the Hunter with his antlers aloft. A freshly broken oak bough took on the appearance of a face, swimming in the gloom above a blended, green tweed coat or the

Rural Work: The Dew Pond on the Height

'country camouflage' of a poacher's many-pocketed jacket. Once, I stood blinking for a long time at a hare, standing tall as a four-foot child in the middle of the field, forelegs hanging unnaturally down by its side, before it dropped to the ground and rocked away; resumed its more readily identifiable shape. Another time, confronted by what I took to be an indistinct roe buck in the near dark, I watched it shape-shift into an upright human form as it stood on its hind legs, swayed and then landed with a gusty bark, before turning tail and becoming deer again. I went with my eyes open, my senses fully alert and zinging. I felt that everything would make sense in one way or another and, increasingly, that I was just another part of it. I wasn't sure what spooked me most though: ghosts, old gods or men. I often had to change direction when I realised the underkeepers were out, shooting. One dark early evening, I had to run for my life along the footpath as shots from a rifle zipped behind and in front of me.

The 'keeper talked about leaving. Reeled me in with his knowledge of how much I loved this place and its wildlife, reminded me of the progress we had made and the certainty that any new gamekeeper wouldn't do as much; would even undo it all. Certainly, no other gamekeeper would let me roam where I wanted. 'Imagine,' he'd say, 'what you'd miss!' He would joke regularly about how often his employer and our landlady would tell him that the place was his, really: that, to all intents and purposes, he had the say. The ownership of it. He would say, often, 'As long as I'm here, the place is yours too. You can go *almost* where you like, as long as I know where you are!' He convinced me I had an input into any conservation plans; a voice. I went out on the nights Martin was home, a little haunted, a little hunted, and with the increasing, creeping feeling this wasn't going to end well. The juxtaposition of our lives shifted. As we drove over soil like broken crockery, the flints rattled, resettled and hardened.

I discovered paths on maps the 'keeper had diverted. Whole swathes of open access across downland and Sites of Special Scientific Interest (SSSIs) that he'd claimed had no access, not even to me, unless I went with him. He fitted padlocks on gates and put up 'Private, Keep Out' signs.

He offered the place to me as if he owned it. And he didn't.

One night, after a silly, tired row at home, I ran out in that strange, crepuscular, post-twilight hour, past the old woods and the Big House and all its ghosts until I couldn't run any more. The melancholy, wavering calls of owls reached me, punctuated with '*ke-wicks*' and their more rarely heard, soft and intimate 'ocarina' warbles. I spotted a tawny owl on a low oak branch, illuminated by the security light of Old Groom's Cottage, a cottage still attached to its own stables and hayloft, although its tenants were away for the working and school week in London. I crept as close as I could to the halo of owl until it regarded me at shed-roof height a mere arm's stretch away.

It did not move or look away. The markings on its chestnut-brown feathers were streaked, spotted and freckled with cream. Its eyes were big dark wells. It turned its head skyward, briefly, the stiff ruff of its facial disc directing and receiving sounds I couldn't hear. For a moment I thought it was going to take off, but instead, it flexed and squeezed its strange, white, brillo-pad toes and talons round its grip of the branch, piercing the crust of lichen and bark. The owl settled its gaze on me with a gathered, renewed intensity. I blinked, being outstared by much bigger, wilder eyes than mine. I had never looked into the eyes of a wild animal for

Rural Work: The Dew Pond on the Height

so long before. To the owl, I was neither predator, usurper or prey. It was an intelligent, penetrating stare. Disquieting. I felt the owl was peering into my soul and assessing it; perhaps finding it wanting.

———

On a cold day in February, 341 years to the week the already-old dew pond on the hill had its name changed from Wigmoreash Pond to Murderer's Pond, and the original gibbet was erected nearby, we find ourselves in it.

We have come, our two families – the 'keeper's and mine, from our villages either side of Inkpen Hill – to restore the ancient dew pond that balances like a saucer on the spine of the rising bascule of chalk, just beyond the long barrow. We are wreathed in cloud. On a good day, the views are far-reaching and panoramic.

Romantic and mysterious, dew ponds are the only source of water up here. There are no springs, rivers or natural ponds. Rainwater sheds quickly off the smooth-domed hills or percolates slowly through chalk to refill aquifers that run the chalk streams in the valleys far below. For a place that was once the sea, it is exceedingly dry.

Goat willow, ash and quickthorn had colonised the pond and were sucking it dry. I had convinced the 'keeper and the owner of the Estate that restoration work was overdue on this historic monument. The top of the lonely path, which skews up Gallows Down and past the pond, had long been a place for summer picnics. Older villagers recalled driving up by pony and trap, with all but the eldest and youngest disembarking at the bottom, to allow the pony to pull up. The beeches, perhaps two hundred years old, are scarred with initials and dates. Natural England advised us that we must cut down the trees growing in the pond but leave the roots in, for fear of breaking up the pan of the basin.

The idea and construction of dew ponds is as old as (if not older than) the hillfort and long barrow that sit upon the same ridge. Gangs of dew-pond makers travelled the country through winter and spring. Four men would take four weeks to make a pond that might last a hundred years or more before it needed maintenance.

A wide, shallow lens, just three feet deep in the centre, was dug out and layered in thick straw, then crushed chalk. A cart and horses would be driven through and over the final layer of chalk, crushing it to a fine powder. Water would be added until it became a thick cream, which was smoothed into a porcelain saucer and left to harden. The dew-pond makers from the deserted village of little Imber on Salisbury Plain would make as many as 15 a year. Before filling naturally, their surface would 'shine like glass'. The filling of them and their mysterious retention of water, even during droughts, has long been debated.

The insulating straw is said to create a thermos effect and initiate dewfall above the surface of the pond. Sometimes a key, overhanging tree was employed to distil water from the very air – they were also known as mist or cloud ponds. It is now thought that the shallow bowl simply retains rainwater, yet the shepherd testifies that water is replenished by morning – even if the night is dry.

When we finished cutting poles, wading and dragging, we bumped along the drove on the tailgate of the Land Rover, splashing through milky puddles. The gibbet appeared and disappeared through shifting mist, and sheep clung to the gorse like low, wet clouds. There is an isolated timelessness to the scene. The creaking of a gate that is no longer there and

Rural Work: The Dew Pond on the Height

the tinkling of sheep bells turn out to be the whistles of red kites, brought low by the weather, and goldfinches chiming through the docks. The sonorous cronks and knocks of wood on wood come unmistakably from the ravens; judging, passing comment, passing sentence.

The date we were there was pure coincidence. I'd wanted to celebrate the older dew pond and its survival down the centuries as Wigmoreash Pond, and not so much its more tragic story of adulterous lovers and a murdered wife and child. I wanted to remake the mirror, as a family. I wanted to return and swim with my husband in it, under the moon.

Days later, after a sudden storm, we went back up to the dew pond. Willow buds on the cut trees had burst, the sap still rising, into pale, lit matches.

In evening sunlight, the little disc reflected back, clear-eyed and brilliant: a lens, a satellite, a mirror to the sky. And when cloud dulled the surface, it was a fallen moon. Skylarks sang upwards and yellowhammers flitted between thorns. Then for a moment, a rainbow arced right over the hill like a banner, its middle lost in cloud, an end in each village. Half-remembered lines from Rudyard Kipling's poem 'Sussex' flitted through my head: 'Our blunt, bow-headed, whale-backed Downs...Half-wild and wholly tame...Only the dewpond on the height/Unfed that never fails'.

A landscape doesn't forget its stories. It wears them like lines on an old face, markings on an old body. There came again, between the 'keeper and I, an increased questioning of motives and truths. The ease between us became strained. And then he made a seismic decision in his personal life that blew both his family and the small community apart. He cut people out. He cut me off, utterly. He reverted to type: gamekeeper and hunter first. Over the next few months, I learnt that he hadn't entirely been who I'd thought he was – who he seemed to present himself as. Perhaps I'd been naïve. Perhaps I needed to look at my own behaviour. Perhaps I'd been told stories. We are all storytellers, after all.

    I was drawn to the hill one wild night. The wind rocked the car and a squall blew in. On an impulse, I got out. No lights were visible in the valley and no one else was there. Yet along the old drove track a mellow, flickering light bobbed. I followed, wondering what it was. Not a torch. Nor someone's phone – not, to my relief, the gamekeeper's lamp. A will-o'-the-wisp then? The phosphorescence of an owl light? It began to rain hard. The light dipped rhythmically as if carried by a hand; a lamp swung backwards and forwards before disappearing near the dew pond. There was the hint of two shapes. I stopped then, wondering quite what I was doing, not having meant to walk that far, realising I'd be missed at home. Although the familiar chalk track gleamed like a pale ribbon, I could not see the fence posts either side. I walked back to the car, started the engine and went to wriggle out of my wet coat before realising that it – and I – were bone dry. The windscreen wipers shuddered, smearing dust, and the lights came back on in the valley far below. Including my own.

CHAPTER FOURTEEN

Watering

Over the next few months, I tried hard not to lose faith. I wondered if the difference I thought I'd made had all been a lie. Could one person make a difference? And had two? Certainly, one person out daily, for half a lifetime, on a 2,500-acre estate, with a tractor, untrammelled quantities of herbicides and pesticides, and the indiscriminate use of flail, mower, strimmer, and next to no knowledge of ecosystems, habitats and even common *species* identification made a difference. So why couldn't I? If it turned out that I hadn't, I wasn't going to give up now. Without ownership in the legal sense, I would have to find other ways to defend and protect nature. Instead of coursing along, half-carried on the slow, lazy, entrenched river I had been on, being *shown* things, being, I suspected now, a little coerced into thinking I'd made a difference, I would have to split into frayed lace tributaries of action that would run through and over and round everything I did or came across. And redouble my efforts.

Throughout the following spring, I went about with one ear half-cocked for the cuckoo, my breath shallow, barely able to look up at the hill and so many places I'd now been excluded and exiled from. And on the day I *did* hear it, its voice seemed to drift, unsettled. I walked towards it, under the down's ruffled fleece of cherry blossom. Its voice seemed chased along by cloud shadows, snagging on thorns at intervals. I realised it was still travelling. My ears and heart strained for it until it dissolved into Wiltshire.

But it was then that I noticed the yellow-green haze on the broad flank of the hill. I'd not been up in a fortnight, sticking (mostly) as I did now to the footpaths, my former freedoms denied. I registered the absence of sheep: grazing sheep are essential to this flower-rich landscape, but in recent consecutive years, they had been allowed to graze the downs indiscriminately. Not enough in some areas and too intensively in others. The rare, rich, fragrant and precious chalk grassland and flowers of this Site of Special Scientific Interest and all its attendant insects and birds had suffered. In some places, rank grasses and nettles had overpowered the low-growing mat of chalk plants, but in the main, the flowers and wild herbs had been grazed off and the insects and birds declined dramatically as a result. I had nagged, cajoled and berated the 'keeper, estate manager and shepherd for some ten years about setting and sticking to a grazing rotation. And the 'keeper had done the same, but to no avail until the previous autumn. A new shepherd had been prevailed upon to keep to a simple grazing rotation, and suddenly, in the evening light, the payoff was revealed: the hill glowed gold. Something had got through. Someone had listened and *acted*. And, as a result, here was a hill full of treasure, nectar and visual delight in the form of millions of tiny, trembling, yellow cowslip bells. It was visible from the bypass.

I tried to visit every day. Caught in late April showers, the cowslips seemed to emanate a buttercup light on the underbellies of low cloud and uplit my chin in a grinning

Watering

selfie I took, even in dull, metallic light. I watched careening hares slipping down through them, down the wet flank of the hill like out-of-control dragsters, all big end over front. I couldn't help but see it as celebratory.

In the days between visits, drifts of birdsong came through like a feverish dream, or a signal that kept dropping out, through the bedroom window, car window, on my way to work or walking the dog. Loud and clear, then indistinct; perhaps, imagined. A grey partridge stopped me in my tracks, then the descending, laughing notes of a willow warbler, a chiffchaff, a starling mimicking the ghosted song of the last lapwing in the village, and then, again, a cuckoo. Songs to shock a heart desperate to hear them, before restarting it into a steadier rhythm. Birds as defibrillators.

On the down in the sun at last, the cowslips formed a sunshine haze. Millions of butter-yellow bells covered the broad expanse, resurgent from their long period of enforced dormancy. They were so thick that it was impossible not to tread on them; to put a hand down in a palm-sized, cowslip-free space. And there would be other treasures to come as the season moved on. Here already was meadow saxifrage, and the anthills, those newly worked castles of fine-sieved soil, had been transformed into blue pillows of speedwell and chalk milkwort. In turn, they would be purple with wild thyme, marjoram and bee orchids and the hill would have a midsummer reprise of gold with a flush of warm-scented lady's bedstraw.

I expected the nectar-filled hill to be instantaneously loud with insects, but it wasn't. I consoled myself that it would take time. That the chalk specialists survived in small colonies on the nearby nature reserve at Ham Hill, that the grizzled skipper, Duke of Burgundy and chalkhill blue butterflies could recolonise this place. There were whitethroats, yellowhammers, meadow pipits and the tented nests of skylarks, but not the remembered key-jangle of corn bunting or the purr of long-gone turtle doves. I stood among the cowslips and desperately wanted the volume back up. The range, subtlety, variation and *loudness* of spring was all diminished. But surely, this was proof that it could be done? Speaking up with a persistent, gentle insistence, shamelessly fluttering my eyelashes and laughing like a willow warbler, I'd played a bit-part in beguiling the cowslips and the bees back to the hill. A softer kind of activism, but an activism, all the same.

That same spring, there was sad news that the last male nightingale on Salisbury Plain had failed to return. Another local extinction. There were swallows in the farmyard below, but just two pairs; not 12 like there were when we first came. When would there be none?

I had lost so many birds since coming here, alone; I stopped to think about my own local extinctions, charted in my local newspaper column: patrolling or 'roding' and nesting woodcock, greenfinch in the garden, lapwing in the lower, village fields, spotted flycatchers, the house martins off the house, where the shadowy cups of their nests remain, like a deserted village. And all the nightingales in Nightingales Wood, behind the house. How can one lose so many birds?

The chimney pot starlings seasonally recorded and broadcast those losses in a stream of mimicry and remembrance. Interrupted by interference and reference points from the human world, they performed a stored library of birdsong, an archive of what we'd lost, interspersed with what we're losing, sometimes gaining, and how we've lost it. The sweet, walloping, yodelling cry of lapwing and the creak of their broader-at-the-tip, owlish wings were still in the starlings'

Watering

repertoire each spring, though they have gone, but the froggy, *'ort, ort ort, squeal'* of the roding woodcock was phased out three years ago. On autumn and spring migration, the piping 'dread' of golden plover contact calls filtered through to my bedroom and I knew they must again be on the hill. Song and mistle thrush had joyfully made a slight return in real life, not just on the starlings' stage, and their borrowed phrases might be followed by the laugh of a green woodpecker, the scream of the toddler down the road, the kestrel's whicker and the underkeeper whistling his dogs. The short bursts of heartstopping and utterly convincing nightingale song were perhaps the biggest wonder. After 30 years of absence here, the song, so strong and memorable, is ghosted, repeated and passed down through consecutive starling generations as a crystal-clear and accurate representation. Starlings generally live for around three years, though the oldest ringed bird was 22 years and 8 months old. Even so, you'd probably have to go way back to listen to a starling that actually got the song from a nightingale here. The chimney pot starling's snatches of lost birdsong are our own criminal record; a soundtrack of guilt and evidence, the score of a shifting baseline syndrome, a spangly black (voice) box recorder, a declining timeline of the sweetest creatures, spiked with pulses of hope. Our haunted, grieving brain fills in the rest.

I have a Patrick Kavanagh quote from an essay called 'The Parish and the Universe', in a collection of essays by the poet, stuck to the wall of the little writing hut my dad bought for me (there is no room to write in the house). The note is pitted with drawing-pin holes and greased with Blu-Tack from being on the walls of rooms I no longer remember. It begins, *'To know fully even one field or one land is a lifetime's experience.'*

I liked it very much. It grounded and earthed me. Justified the fierceness of a feeling that I never wanted to be away from here – at least, not for very long. Yet, in the process of being exiled from much of my former free range, I found I must.

I'd written a book on otters, a natural history for the RSPB in their 'Spotlight' series. In writing it, I'd discovered the need to find water, away from the paradox of what Kipling called in his poem 'Sussex', on the South Downs, this dry, 'broad and brookless' place, that looked like a sea, was once a sea and that percolates, filters, stores and releases water, but doesn't keep it for itself. William Cobbett remarked on this in his *Rural Rides*, in 1830: 'To stand upon any of the hills and look around you, you almost think you see the ups and downs of sea in a heavy swell…in a country so full of hills one would expect endless runs of water and springs. There are none: absolutely none. No water-furrow is ever made in the land. No ditches round the fields.' I went back down to the chalk streams: the Kennet, Dun and Lambourn, the Enborne, Pang, Anton, Test and Itchen. The clear, benign and shallow rivers and streams that run at an almost constant temperature of 10°C are renowned for their fishing. As such, of course, for most of their lengths, they are almost all privately owned and inaccessible to most.

When I had been writing about otters, it had become all-consuming. I tracked otters, read about otters and dreamt, every night, about otters. They are tricksy, mercurial beasts; ribbons of water, ribbons of fur, nocturnal, acutely people-shy, vanishy – and then just as likely to turn up like clockwork in a town-centre river on Market Day for a fortnight. For those months, I felt as if an otter was curled up wetly in my head like a strange hat, river water dribbling through the runnels of my brain and leaking out of my eyes, to drip over my eyelashes.

Watering

I could almost feel its drying fur turning spiky, opening up the pelt to dry. The glimpses and near misses I had were utterly bewitching. I longed for whatever they would give me: a fresh spraint smelling of roll-mop herrings, ashtrays and new-mown hay. Half a pawprint, rayed like the sun behind a half-cloud; a wet trail drying on the rocks of an oxbow 'island' in the Scottish mountains; or the glimpse of a Labrador-thick tail, turning like a thrown stick in the broil of a Hungerford weir, then gone before I registered what it was. I honed my skills. My head was a holt, a couch, a hover for otters.

I saw them eventually, stopping at every river I passed, towing my husband and children along at any opportunity. I saw their broad, border-terrier heads swimming in the River Dun beside me, on the parapets of Hungerford Bridge in the evening, beside the Co-Op garage, beneath the rumble of cars and tractors crossing the town bridge – or swimming low, indistinct and serpentine through a pool of orange sodium light in Andover Town Centre. Trespassing lightly one evening, sat on the end of a low wooden bridge, I was approached by a muntjac deer. It came closer and closer until it put its face up to mine and we bumped noses. Realising what I was, it jumped back and wheeled round in surprise, lost its footing and landed with a splash in the water. Evidently a strong swimmer, it struck out and scrambled up the bank, quickly disappearing into the tall vegetation with its white tail waving like a flag of both surrender and alarm. Not minutes later, I had followed it in: unable to place the high, rising whine of the railway tracks singing to the tune of the approaching evening train to Exeter (a grasshopper warbler? No…). I jumped out of my skin when it roared past, forgotten metres away. I lost my balance and fell in, feet first. Although the water was just knee-deep, my wellies filled and the scramble up the path the muntjac had made through the reeds was a difficult and ignominious one.

I sought that intensity again, just a very little away from home. In searching for otters, I'd gone searching for water, starting at the source of one of the chalk stream tributaries of the River Kennet: the Enborne, which began within the park pale of the Big House, opposite its Queen Anne frontage, beyond its topiary birds and lawn of spring crocuses. This time, I'd begun tracing it from the top of the hill on a day of pouring rain, with the chalky water running like spilt milk it was no use crying over, down the road. The further I went, the clearer it got, cross-hatching the camber in sinuous ribbons, a chain of Tarka-bubbles keeping pace, as though there were otters just beneath the surface.

I walked away, right off the Estate and discovered other places. Discovered that I existed there too. One late February evening, on the trail of a locally made YouTube clip of a starling murmuration over Speen Moors, I walked down a footpath familiar to me 20-odd years ago, shouldering my binoculars. Near this spot, the three Newbury Martyrs, including Julins Palmer, had been burnt for having the 'wrong' religious beliefs in 1556; near this spot, Royalist and Roundhead soldiers had massed, taken positions and fought, countryman against countryman, for their lives; one of those Parliamentarians, London cheesemonger Thomas Prince, was wounded on the hill opposite, but survived to become a founder member of the Leveller movement, along with six others (including two women). They fought for a nascent democracy. For equality, religious freedom, press freedoms and the freedom of land for all, as put forward in their manifesto, 'a peace offering to this distressed nation'. Near this spot, the Speenhamland System (or Berkshire Bread

Watering

Act) was born to try and alleviate rural poverty and prevent discontent. Near this spot, William Cobbett had spoken directly to and roused the poor, landless and disenfranchised agricultural labourers. Near this spot, I had fought my own environmental battle.

It was after sunset and I was wrong-footed, immediately. There were lights moving through the trees and the sound of what I first took to be birds and water was the roar of traffic. I was in the wrong place and found myself walking parallel with the bypass I had fought to prevent more than two decades ago. My nerves were already frayed by the noise and darkness and I felt a little undone. I pressed on.

The senses I relied upon for walking alone in dark places were compromised by the road. I stepped down onto a pale, concrete track and thought for a falling, dizzying moment it was the river. I felt like I might cry with a mixture of fright, a sense of my own perceived stupidity, going somewhere like this in the dark, and what the place had become. But the possibility of seeing tens of thousands of starlings compelled me on. Abruptly, where there should have been a chalk stream, the path ended in a dark rectangle of stagnant water that glinted and smelt like oil. I couldn't make sense of it at first. I tried to calm my breathing and take stock. This was a flooded underpass.

Bats looped through clouds of early season midges and water dripped loudly in the gaps between traffic, making me jump. I was not ready to turn back yet. As far as I could see there was a concrete ledge. I made a leap in the dark and clung as best I could to the smooth concrete wall and shuffled towards the pale square on the other side.

The light had gone from the sky and so had any starlings. But the landscape of Speen Moors opened up, familiar and welcoming. My memory sharpened with the first stars. I recognised the filigree of alders, the rickety bridge and the flint-coloured chalk stream I once rowed across in a borrowed coracle using my hands as paddles. I should have been home an hour ago. Eventually, I retraced my somewhat light and fearful steps, my senses heightened. A road sign high above my head, criss-crossed with winter branches, masqueraded as a ghostly tree camp; a rattling plastic bag, a fluttering flag. And then my imagination was in overdrive. I thought I could see vertical movement towards the river, against the flow of traffic: people in lobster-tailed helmets and white gauntlets; the ash poles of the trees became like pikes, lowered and bristling at the road.

Fear, perhaps, breeds fancy; landscapes, perhaps, can be like great sponges, taking up violent, highly charged human activities as they do water, playing them back like rolling mists over flooded, and reflooded, taken and retaken ground. What business had this road to be here?

And then another strange, but more tangible thing happened. I was hyper alert – to what I wasn't sure. Invisible threats, strangers, ghosts? My senses were wound to their height. I could see better now. I could smell petrol, but above that, cherry laurel blossom from the old keepered woods, a fox, a dead animal and stagnant water. And, impossible but true, the very late evensong of birds was enhanced loudly, over all the road noise. Song and mistle thrush, dunnock, robin and a wren so loud that it felt like they might burst my eardrums. And then, the high whistle of an otter. I shouldn't have been able to hear any of this above the traffic noise, but I could. Beautiful, clear and powerfully felt, in the almost dark.

I felt raw, but stronger. Like Proserpina emerging from six months in the underworld into a warm spring evening,

Watering

where the blackbirds sang past dusk. I wondered at other ways to protest; to save things, to shout above the noise for nature, wren-like, thrush-like, otter-like and, in between, perform little opportune acts of activism – I knew this was through writing. Reading and writing as a form of protest and galvanising action.

Back home, I made transgressions. Small emotional trespasses back onto the land, using and refining my female, animal senses and, using my own key, I let myself back in. One evening, our landlady's horses got loose in the park and galloped around. I went out with a headcollar swinging from my shoulder and stepped around the swallow holes where the water percolated down and came up again like stitched needle holes across the grass – potential fetlock snappers. The horses were caught by the housekeeper and I was left looking at the spring's source running through the lake and away into what becomes the River Enborne. There were rumours that otters had been seen in the lake of this house, and in that of the neighbouring estate, too. Exploratory diversions. I smiled wryly, thinking that otters might have travelled upstream to the source, too.

I wrote with a mustelid hunger. Like a dam that would burst if not released, becoming necessarily otter-like in my writing and walking. Snatching at moments, going with the flow, elusive but predictable along my own home stretch, camouflaged and never on time; fossicking about, diversifying habits, hours and timings in response to weather, water tables, the movements of humans and the world out there – family.

With so little action taken for the environment, for wildlife in the intervening years since the bypass was built, there had become a renewed urgency to it. The time, not so much now, but passing. Lines of Seamus Heaney's that so struck me in my sixth-form and university days came back with fresh power to haunt me: how he, as a poet, felt a responsibility to write – to write the wrongs – but at the same time wondered whether all he was doing was blowing up sparks. In a newspaper cutting I had tucked into my copy of his poetry collection *North*, from an interview he did with fellow Irish poet Dennis O'Driscoll, in *The Guardian* in 2008, Heaney said that 'defiance is actually part of the lyric job'. I'd circled the line, along with others. And whenever, sporadically, I reached for the book or poem, there was that line. To a greater or lesser extent, writing about nature had always dealt with loss. But now it had long passed the point of being imperative. It is impossible to write with integrity about nature without protesting and resisting and waving a desperate red flag.

Isn't it?

We are writing for our very lives and for those wild lives we share this one, lonely planet with.

Writing was also a way to channel the wildness; to investigate and interpret it, to give it a voice and defend it. But it was also a connection between home and action; a plank bridge between a domestic and wild sense. A way both to home *and* to resist.

The children grew up, as children do. Billy turned 17 and left school for a music course at Andover College, some 13 miles away over the hill. He learnt to drive a car (a necessary evil when there was no bus to get him there in the morning and a return bus that dropped him seven miles from the door). A text came from him one day, while I was at work in the library; 'You'll *never* guess what I'm having lunch with!' There

Watering

followed the most exquisite video clip, taken on his phone. An otter, hunting the shallow stream at the entrance to the music block, rolling, roiling and revelling in its element, stirring up chalk sediment, a crisp packet and riffling the thickening mat of watercress bobbing on the surface. In broad daylight, between a busy roundabout, car parks and classrooms.

He'd walked right past it at first, glimpsing it before he realised what it was and stopped, in the midst of his long, countryman's lope, and walked backwards. Otter.

He watched it move under the gin-clear chalk stream, fluent as a water snake, spiralling and turning on itself with the ease of a gymnast's ribbon, writing itself in cursive flourishes through the water. He glimpsed a paw; a hind foot kicking off, a thick rudder tail working hard in a tight spot. He described how it swam through the pipe-tunnel of a road bridge and went up the bank: water thickening into fur, dissolving back into water. He saw the champagne belt of bubbles, fizzing in its wake, a bubbling galaxy of stars exploding in the dark water behind it.

He enjoyed the spectacle for as long as he could. An unlikely wildlife watcher, perhaps: bass guitar in hand, leather jacket and Dr Martens on, a lad in need of a haircut, grinning, texting his mum.

It wasn't his first otter, but it was most definitely his best. And my goodness he'd earned it. It was his reward and his alone. A naturalist's dream of a 'walkaway view'. He left and went to band practice.

Later, we watched his joyful video clip go viral on Twitter. I went outside in the dark to my little, bow-windowed hut, and wrote about it. It was a wild night, with the full force of a southwesterly coming off the hill and hitting the prow of the hut like a small, storm-tossed boat. A full ten yards from the lit house, I could have been miles away at sea, unmoored, riding out a storm with the rain finding its way through the window joints, soaking my notebook and spitting in my face. And then I thought that what remains is that it matters, to record all this. It matters to delight in nature, and sometimes, despair of the losses; it matters to set down words – along with other people that write about, shout about, reason and mediate about nature. It's vital. Deeds *from* words, perhaps. Because what will survive of us is love and, possibly, the evidence of the printed word. And in that love recorded, there is a kind of ecology.

One night, when the tap in the yard had frozen, I took the buckets to the spring in the park and dipped them in. Frost sugared the grass. A snipe jinked up with the sound of tearing fabric, as if it had been frozen to the earth. To each horse, I carried the moon in a bucket.

CHAPTER FIFTEEN

An Ecology of Love and Ruin

I marked our seasons with diminishing birdsong, our love by the year. When each quarter came around again, there was an anxious stockchecking of what remained and what had been lost. All across the countryside, the same battles cycled round, interspersed with new ones: field headlands mowed needlessly, hedges removed in bird-nesting time and winter hedges decimated several times annually by the flail, so they were never allowed to flower, denying nectar and cover and spilling all the livid berries onto the road like broken necklaces. All that bright sustenance crushed by passing cars and the birds of the air gone hungry. And there were nasty surprises, too – birds clumsily netted out of hedgerows that might be affected by new developments: nets that entangled a cock blackbird, a hen tree sparrow, a thrush, a hedgehog, an owl. And further afield, news of the famously leafy city of Sheffield robbed of its heartwood, its street trees felled; a rail link ripping through ancient, irreplaceable habitat and people up trees again to save them. All for the purpose of what exactly? To save money in the short term? To shave minutes off a commute?

Could we not be more creative than that? Overseas and in our supermarkets and cupboards, the incomprehensible horror of nocturnal, mechanical olive harvests that sucked up countless roosting songbirds – nightingales, willow warblers, song thrushes, blackbirds – vacuumed to death as bycatch. New roads, more plastic, fracking, mining, burning and all the other ways we have thought of to burn our own house down, conspired to make us feel helpless; sucked into the enormity of how we live now. The climate emergency and the loss of habitat and the draining away of our wildlife seems desperately overwhelming.

But all my best memories, of love and family and living, have been spent outdoors in nature. How can we stop fighting for this? I've 'hefted' my children here with wild words, stories and natural artefacts, landmarks. I wanted the place, its meanings, histories and connections to keep happening; to continue and to emanate a strong pull for them. Magnetic home. But, as they got older, I didn't want to hold them here. I wanted to make sure that, wherever they ended up, the nature of the place they found themselves in would be a talisman for them; a lodestone, a waymarker, a way in and a way to 'home'.

I twined it all up for them, making a magic that would hold them to this place which was safe, protected and protecting – but perhaps not in the usual sense of those words. Because it also needed protecting; it was a two-way relationship with a place that could only come from the gentle and persistent enquiry of exploration and connection. It was important to understand that actually, often, this had little to do with 'coming from' this place; that the sense of belonging could only come from an engagement with it – and that this was portable, a passport in fact. The freedom to make their own connections, write their own stories and create their own rituals. I wanted them to have the experience of those birds that migrate here, that know two places (and several more) in the world, intimately. Nature offered me a key to all my houses and hopefully would to theirs, too.

Every second or third year, we'd go about in spring with a box of pearls in our coat pockets; little green clasps attached to their tops. Though it didn't seem the season, these milky, translucent, sticky spheres (like trout eggs) were mistletoe berries, sometimes saved in a Tupperware box in the freezer since Christmas. We sought out its host species around the estate, and a little beyond it; crab apple, poplar, hawthorn, lime, oak (though oak mistletoe is rare) and 'planted' them. I loved our silly family tradition of it, with its air of trespass and naughtiness, and of course, the reverence for the magic and symbolism of the plant, which seemed especially pertinent to us. I liked the fact that it hung in the air, ungrounded, yet belonging, hidden in plain sight: the knotted green baubles only revealed in certain lights or seasons. Bird-sown by a beak wipe, or defecation, mistletoe is absolutely of a place, but owns no ground and has no roots of its own. It is a borrower. A symbiotic part, a tenant. I liked the way that perhaps, when we were long gone, there might be traces of us in the landscape; globes in the trees, golden-green sky baubles.

I'd made a habit of 'planting' some wherever we'd lived. It didn't always take, but a misty globe hovers, netted in the branches of a now-tall poplar tree at the foot of my last childhood home at Wash Water. I didn't need to plant any at Highclere, but had loved pushing the pram down the lime avenue to the Castle, where great balls of it hung like lampshades grouped in an art installation. One tree alone sported 40 ethereal balls. One of these balls measuring four feet across.

The best way to propagate mistletoe is by wiping it on a young branch. We popped the pearls for the hard, heart-shaped, frog-hopper-sized seed inside and stuck them to a branch with their own viscous, pearlised adhesive. If they survive to germinate, bend, make contact with the cambium of their host (and not get eaten by birds or insects), in about five years there may appear a sudden golden-green wishbone and a bright symmetry of paired leaves, like an aerial ash tree seedling. Quickening then, over the years, the fine, wispy knots and pliable, brassy twigs knot and repeat in complicated Celtic patterns. And there may be new semi-opaque pearls from this ancient plant – and excuses for kisses, all over the place. It felt like a way to travel.

I hadn't owned a passport or been out of the country for around 17 years; Martin's expired a little after mine. There were times when this made us feel like non-citizens, trying to prove we existed, at our age, without a mortgage or a passport. Only two of our children have had one and only for a residential school trip to France and Germany. None of them had ever been on a plane (unless you count the rare thrill of a silent, kite-like glider, soaring over the hill and house from a downland airfield with the Scouts). For all those years, we couldn't afford to go anywhere, so there was little point in passports, and then the cost of five was completely unreachable. Our two teens in particular were desperate to explore the world – and will – and when they do, they will know how precious this freedom and privilege is and will do so carefully and as responsibly as possible. With a sensitivity for an intimacy with the world and its other inhabitants, home might be anywhere. Through nature, I hope they will have a way in. Growing up in such a small place, there will always be an 'out there': many 'out theres' and aways.

An Ecology of Love and Ruin

But there will always be a 'here'. Even if we or the house are gone, finally replaced by one three times its size, there will always be Gallows Down. The centre they have spun out from, the middle of the map. A giantess lying on her side, the combe of her waist and height of her bone-chalk, thin-skinned hips and the gathering of her dark tresses in the trees of the hanging woods. Then, under the vastness of the dark, starlit night above, neither hill nor sky, nor its reaches back through time to when it was an ocean, could ever be considered small.

The children have their own elemental place in the order of things. They are my compass points, my bearings, my northings and eastings – all points home – and all the birds of the air fly between them. I hope that, above all, I have given them roots and wings. And the confidence to make their own stories.

―――

The place and our wild moments in it are carried within us, just as the water we drink sheds off these hills or percolates through, is stored in aquifers until it courses along the chalk streams, is taken (too much), treated, and returned through our taps. Here is where the chalk resurfaces, furring up the kettle and lagging the washing machine (turn that dial to hear wintering fieldfares, '*chack, chack, chack*'!), or accumulating on the element of the dishwasher (see those fork tines tangle like antlers?), calcifying around the handles of our electric toothbrushes (the one that doesn't work so well, and runs itself out against the cup, with the precise, mechanical notes of a nightjar), or sparkling through the milky dew pond of a cereal bowl, left on the gatepost as the school bus pulls up (the sound of its door opening, that exact call of the barn owl that flies over at dusk, or more so, that of my son's Dr Who Sonic Screwdriver, lying like a museum piece on his shelf). The chalk re-enters me with its particular tang in a cup of tea, cooling beside my laptop as I write (the scroll key on the mouse, a raven's *'broaak, broaak, broaak'* down the length of the written page). Each a cinematic, synaptic flicker of connection, recalling fresh, bright moments with clarity and pricking instincts to recall those moments when you felt most alive.

But then suddenly, unexpectedly, Dad became ill and died within a month. He and mum had moved to a smallholding in North Devon when I was expecting Billy and had more than 15 years there, in a white-washed Devon longhouse that came with a goat, raising a few cows, sheep, ducks and hens. They'd moved back to be near us in the village, both fit and well, active and independent, about three years ago. Dad was 76. He and Mum had had the holiday of a lifetime in the January, visiting my dear brother and family who lived in Australia. Towards the end of the summer, Reading Festival time, Dad had begun to seem a little vacant, a little lost. He slept a lot, lost interest in everything, and we wondered about depression. He baffled the doctors. And then he worsened and worsened, became confused and lost the ability to walk. They kept him in hospital, at the Royal Berks in Reading, 30 miles away, where our youngest, Rosie, had been born. The hospital that is opposite the Victorian red-brick Museum of English Rural Life, or 'The MERL' as it is known with great affection. I took comfort from that and some solace in visiting its much-loved and familiar wagons, carts and ploughs, its billhooks, scythes, smocks and baskets, and smiled wryly at a glass case exhibiting the particular clothes worn by Jim Hindle, fellow protestor on the bypass and author of a book about his time there, *Nine Miles*. In the opposite bed, an elderly gentleman regaled Mum with fond memories of his working years – at a sawmill in a village right 'out in the sticks, with tumbledown thatched cottages and a big hill.' What were the chances? We shared the same village, the same lane connecting us. From one of the hospital windows, you could see the hill, navy in the distance, a bow-headed whale surfacing.

And then at last, we had a diagnosis.

My brother, who had been over for a visit in between jobs, broke his heart saying goodbye. Then, with my brave, independent and spirited mum and my chalk hill and linnet song of a husband, we broke Dad out of hospital and got him home. Mum cared for him with immense love, strength, patience and care, with Martin's unswervingly selfless, practical and personal professionalism, and I did what I could to support them. Mum and Dad's front door faced the skew path up the hill at a few fields' remove. And when I could, I ran up there; the golden light and air of September so soft, I wrapped it around me. I ran, rode the horse and walked. Up the ancient chalk and flint holloway of Bunjum Lane, towards Prosperous Farm, where Jethro Tull invented the seed drill, some puzzling chalkpit graffiti appeared: an old name scrawled on a new fence in an old, old place. Above a storm drain, dug into an old cottage-garden rubbish dump of Shippam's Paste bottles, dented tin teapots and shards of thick, ship's-biscuit earthenware, 'Oliver Cromwell' was written in looping, cursive hand. I wondered if old allegiances died hard out here, just a short afternoon's walk from the Civil War battle site of 1643, in these times of political turmoil. Along what my Romany grandad would call the 'pobbel drom', crab apples beaded and perfumed the tarmac with a rich seam of golden green. Nuts and berries cracked underfoot and, where they had been milled by tyres, hooves and booted feet, made a pale green flour.

I first noticed the dead hare when I stepped over the stile and disturbed three kites, two ravens and a buzzard. The poor carcass was still fresh and intact. I retreated and the birds organised themselves into a counsel of waiting.

Below the big hill, a Venn diagram of circles was scored into the stubble; the doughnuts of a poacher's vehicle chasing a dog, chasing a hare.

In the four short weeks it took to lose Dad, just 17 days after a diagnosis of an aggressive brain tumour, I marked the decomposition of the hare. There was no obvious mark on her, but even in death she had been harried into a fleeing position by the birds. Her body opened to the elements. A small hole just behind her heart began to seethe with fly larvae. A red-chequered sexton beetle investigated and a devil's coach horse beetle ran over the power-housing of her haunches. The breeze stroked her soft fur the wrong way. She remained elegant as a thoroughbred.

I walked up again and again. Below the big hill, there was a smoke haze of bonfires where only last month there were plumes of chalk and chaff dust from combine harvesters. I could taste the metallic tang of ash in the air as the first fieldfares came in. Back at Mum and Dad's, a tawny owl called outside the house in daytime, as it had for the past four weeks. And when I got home, one was calling at home too, at three in the afternoon, in the golden straw-light of post-harvest. I knew what it meant and I didn't need telling. Dad died the following morning at his and Mum's home. I bore the owls no grudge; I wasn't really superstitious and it remained a sound that soothed and wilded my nights.

An Ecology of Love and Ruin

They continued calling into the next day, when Mum and I both came home to village flowers heaped on the doorstep. Owls and flowers.

I walked for a whole day on the hill and sat for hours hugging my knees and my dog, with views out to almost all the houses I've ever lived in. An early attempt at a cure for grief. Pale, bleached grasses, pewter skies, leaves thin as watery light. This was a spent and exhausted countryside, beginning to put itself to bed. I took strength from the thin turf and light, gleaned kernels, let husks blow away.

At the autumn equinox, I tried to balance the light with the dark as the last swallows filtered through, tilting at kites and ravens and the stillness of hares in their forms.

Four weeks and a day after finding her, the hare's skeletal form was revealed and she was left alone; her skullcap of black leathery ears pricked forward. Her well-sprung ribs exposed like a whale's and fanned-out like egret feathers. Her spine, serpentine. Her hind legs, like train pistons; the long thigh, shin and foot bones at right angles, powering the dragster cogs of a body built for speed, to who knows where.

My last visit was at night. The empty dew-pond socket of her eye gazed blankly at faint stars. The moon gleamed on white-chalk bones, long white claws and the tufts of pale fur snagged at her fetlocks, like the winged feet of Hermes. An appropriate god perhaps; returning her to the earth she will always squat in. Her last form.

Outside the window, there were towering walls of gold. Castellated fortresses of straw bales, bigger than our house, to soften floors and stop up bitter draughts under stable doors and between barn walls, when a gold-threaded memory of warm summer is most needed. Long yellow tresses of straw hung from the hedges and low trees like bunting, festooning and banking the road edges and blowing through the door into the house. A countryside en fête. A crow called in a surprisingly strong West Berkshire accent, '*Oi! Boi! Byre, fyre!*' and a raven called like a stone gone down a well. Farmer Carter, then Farmer Stokes, stopped on the lane to say: 'We heard along the grapevine and are sorry.' But I think the corvids told. There are no grapevines out here. Only bedwine. Bedwine, woodbine, smokewynd, traveller's joy, wild clematis. It has so many names, but bedwine, a corruption of 'between' and 'entwined', and said in the local accent, where the 'd' is dropped and the 'i' becomes a 'y', summons its between-worlds, between-seasons continuity, twining like a rope way I can hold on to, away from home and back again. It has a phantom quality. Its winter smoke-blossom has an ability to appear and disappear, seemingly at will – though this is weather related. In wet conditions, the sodden, damson-coloured seedheads turn spidery and go unnoticed, but in dry, the feathery plumes, like library-laddered moth antennae, fluff and curl in on themselves to form globes that catch the light. Then, they can form ethereal silver, gold or bronze lights, appearing at the same hour for days before disappearing altogether, smoking through the hedges like a remnant, gone-out wildfire.

On the day we went to officially register Dad's death, we were also printing off slogans to save the planet. In her GCSE year, our middle child, Evie, had written an impassioned, clear-eyed letter to her Headteacher, explaining why she could not come to school that day. She got the Head's blessing. Along with her best friend and Head Girl, Ceara, they were striking for the climate in the nearby, bigger town of Newbury, some 23 years since I protested there. We joined around three hundred others on a day when adults were asked to strike alongside the children and young adults. The sixth-formers from the two larger Newbury schools led the way. They were organised, confident, eloquent, informed and impassioned. They meant this. We followed their lead, marched and chanted. Outside the council offices where I had protested against the Poll Tax, the bypass, the loss of rural bus services and the devastating cuts made to our essential library service (that cost many colleagues their jobs), we listened to their speeches and admired their witty, engaging placards. We had brought our own, too. Mine said, 'I was here 23 years ago. I'm here with my daughter today. We need change now.' Colourful, stylish Extinction Rebellion flags fluttered in the breeze. Their flapping and snapping took me back to those flags at the tops of trees on the bypass. These young people will go on to great things, if there's a planet left to stand on and air to breathe. The atmosphere was good-natured, supportive and very emotional. Mum had come too, just two days after losing her husband of 52 years. She stepped out, listened from a distance outside the council offices where she'd worked before she and Dad retired and moved to Devon, and I wondered what I was thinking of, coming here, bringing Mum, just before we had to formally register Dad's death? There were many people here that I knew, that we knew. I had some strong hugs. Wordless, heartfelt hand squeezes. Evie recognised old friends from Scouts, there were some I went to school with and people I was on the bypass with. It felt absolutely important and right to be there – but also, I wondered how we were even functioning.

 I'd helped the girls form an Environment Club at school, where it was desperately needed. Theirs was a tough audience. Our three children had developed natural antennae for counter-culture. Partly fuelled, perhaps, by growing up outside much of popular culture. Living in the sticks where money was constantly tight, much of this wasn't accessible and they learnt to be more resourceful, reflective and considered. To be on the side of the underdog. By this time, Billy was playing bass guitar in three different punk bands – writing songs with his mates from college about the fragility of men's mental health, feminism, politics and, of course, the environment; shouting above the noise.

Though I doubted the wisdom of it and worried about Mum, it felt absolutely right for me to be marching. Dad would have scoffed. Rolled his eyes. And made us the best banner. He'd wonder why it all meant so much, too much – and I'd have told him this. That nature is everything. It is the place I come from and the place I go to. It is family. It is memory. It is a going forwards, an exploration. Wherever I am, it is home and away, an escape, a bolthole, a reason, a place to fight for, a consolation, and a way home.

CHAPTER SIXTEEN

Kite Flying

The spring before we lost Dad, the wood the children had grown up in (and countless village children before them) was ravaged, shockingly, unexpectedly and without warning, for its timber. The wood, known by several names, but to us as 'The Sticks Walk' was at the centre of our strung-out village. To us all and for generations, it was a free-roaming, open-access place of dens, dams and bridges, dog walkers, horse riders, mountain bikers, badger and bird watchers. The rich bright ochre clay by the Ingle brook, which my children called the 'Treacle Mine', had in the past sustained a small pottery industry and earned villagers the old name of 'yellowlegs', their boots and calves no doubt always splashed and smeared by the sunset-coloured mud. It resurfaced still – the mud and the name – all over school socks and pram wheels, and rubbed off on brass-band and rock-band names, the women's book club (The Yellow Pages) and the men's drinking club. Old RSPB magazines are full of pictures of us in the wood.

Historically part of Inkpen's 'Great Common', much of it was mixed, deciduous woodland, with streams, remnant patches of derelict heathland, pioneer self-seeded birch and pine, and an old, forgotten plantation of fir trees. Except, someone hadn't forgotten. One of the entrances to the wood was cathedralled and stain-glassed with a spectacular and much-loved avenue of strong, healthy and towering beech trees. They made great, arcing flying buttresses and vegetal, fan-vaulted roofs over the lane. We loved the green underwatery light they made on the road to school. It was one of my son's favourite places to cycle under. One spring morning, in the middle of bird-nesting season, we came home from school to find all the beeches felled on one side. Along with oaks and ash of a substantial age, they were already being stacked. It was as if a bomb had hit the aisle of our cathedral. Nobody knew the men wielding the chainsaws and arrogantly blocking the road and the bus routes, taking their measured time. They were evasive and rude, knew nothing nor cared anything about the place, its past, present, future and our relationship with it. They told everyone the trees were dangerous. That they'd spoken to the immediate neighbours (the two houses closest to) and would do no more. And they'd replant the trees.

The following day, on the way to school, my daughter and I walked down the pot-holed road to meet them. Me, in my heels and best Lois Lane trench coat, with a slick of postbox-red lipstick, a smile and a reporter's notebook, my smallest daughter suppressing all her fury in her school uniform and waist-length, wild and wavy hair.

The summer migrant birds had begun to arrive. I had already heard chiffchaff and willow warbler and seen the first swallows over the farmyard. But as we walked that morning, the cuckoo began calling from the very wood that was being felled. My daughter and I stopped in our tracks to listen. She smiled a big broad smile at me, but I had to bite my lip to stop tears of anger pricking at my eyes. The cuckoo was almost always heard here first in the village. But what was it returning to? Surely it wouldn't stay – and then, would that be it? Would that be the last cuckoo heard in the village? My phone pinged with a text message from my friend Jules on the other side of the wood, 'I can hear the cuckoo, can you?' I thought of all the people since time immemorial that would have marked this arrival, this return – perhaps with a nod of acknowledgement towards the bird itself, or to a neighbour; perhaps in a conversation outside a cottage door or between a Lord of the Manor to his groom or the thatcher up on the roof to a woman hanging out washing below. The return of the cuckoo would always have been marked and remarked upon by someone. I took a deep breath, retied the belt of my coat and stepped forward with fresh resolve.

My phone pinged again with a message from my other daughter, Evie, who had just gone by on the school mini-bus, the sight of the felled trees causing the bus load of children to erupt in anger as they went past, Instagramming pictures on their phones in turn. Boudicca's children, rebelling.

I reached the men and smiled. Asked if I could trouble the foreman a moment, as I understood some of the villagers had been giving them grief? Gave him my pseudonym, Captain Swing-style, borrowed from long-gone, dispossessed commoners and my mother's maiden name. I feigned indifference and gave him some conspiratorial confidence, faked a newspaper story and got stuff out of him. 'Between you, me and your newspaper,' he joked, 'who do these locals think they are? They don't own this wood! I don't give a stuff if they think they do. We're going to take a lot more trees down through the wood, then fence off this footpath so they can't ever use it again. That'll teach them.' I asked if the beech trees were really dangerous. 'All trees are, when they need to be,' he smirked. I'd got enough out of him. 'And where is the ash tree, with the protected kite's nest in it?' I asked. He hesitated. Hardened. Reddened. 'No birds nesting here, love,' he said. 'And no ash trees. These are all old beech. Some oak.' He leaned down, loomed. Then planted his legs deliberately, proprietorially. Hands in pockets, chest and groin thrust out, he made himself as large as possible. I smiled again. 'It's bird-nesting season. That's an ash tree. And so is that,' I said, patting them. I could tell oak from ash, chainsaw from handsaw, even with all their limbs lopped off. 'Have you done an environmental impact assessment? A wildlife survey? There are rare fungi here, dormice, a badger sett, water shrews, adders.'

Protected species, undocumented. I couldn't produce papers or reports or an inventory of formerly free land to show the foresters, like a magic trick; a receipt for a place, its access and now its things, carelessly, *particularly* lost, because itemising them as evidence or collateral hadn't mattered before.

I felt small and silly and ineffectual in my anger, in front of the men with my daughter. But she stood tall and primed for action, incensed and righteous, a whippy sapling rooted in the place and confident in her conviction that this would not happen. But I had to turn away. I, who had brought her up with such stories. Of how I lay in front of cherry pickers and breached police lines, stood in front of men wielding chainsaws inches from my face and wrists in another wood. But what could I do here?

He leaned in close to me then, put a heavy hand on my shoulder and said, deliberately and slowly, 'You take care miss, won't you. I've been here before. You take good care of yourself now.' I smiled my thanks for his concern, showed no sign of acknowledging this thinly veiled threat and walked away with my daughter, holding her hand a little too tightly. 'So have I,' I muttered, 'I've been here before too.'

The footpaths were our teenagers' independent routes out of here – and back home again: tracks worn by so many feet to the train station in the next village, off the pavement-less, unlit roads. Safe, scenic, secret. They were not going to give this up.

The village parish council got together, led by our steely, indomitable clerk. It seemed hopeless. The wood *was* privately owned, the forestry work legitimate. But the men had not reckoned on a community shocked by austerity cuts and galvanised by a deeply shared history of childhood in the woods.

We were angry at the way it was done. The subterfuge, the sudden arrival no one could prepare for. The legal flouting of what we all believed was law. The century-old paths had been deliberately, cynically targeted, as had favourite trees, rope swings and dens. The scent of rising, leaking sap, split greenwood and mashed foliage mingled potently with petrol. I searched mournfully among the wreckage for bird and dormice nests. For evidence.

We women met over cups of tea, chocolate brownies and pear bellinis. While the men of the Yellow Legs Drinking Club talked about us with fond 'patrony', urged caution and compared lawnmower sizes in the pub, we worked. We used social media, we connected and gained confidence and leads. From 170 straight miles up the country, a line is thrown from the indomitable Sheffield Tree Action Group defenders and protestors. Someone knows of our man. He is a respected local businessman in the West Country, leading a campaign to stop destruction of woodland around his own home, for some of the exact same reasons we were. I rang him, softening my voice, even as he raised his. He thinks us rude, vile even. It is his wood now. They are his trees. But he will do a wildlife survey (he didn't).

In our woods, we moved their things, remade paths, reworded their 'Keep Out' signs with politic politeness and drew the same old lines on new maps.

And in the end, the power of our collective memory and shared stories partially won. My five-year-old records of dormice were too old to count. Our anecdotes on other species, just that. Anecdotal. Unverified. But just as the tenderest leaves were unfurling the following year and they came to 'finish the job', trashing it anew, it looked like we might be granted our legal footpaths. The cuckoo returned and called fitfully that year, not seeming to stay for long in any one place. Settling eventually for a few weeks near the badger sett.

One late summer afternoon, I was despairing about the state of the house. How we'd long outgrown it, how the fence between us and both neighbours was falling down, how the underkeeper's dog got through and burst into our house when the girls came home from school, attacking our own dog on her bed in our living room and then turning on the girls. How our damp, tiny, 1 × 2-metre shower room had gone beyond scrubbing and regular repainting when we found the linoleum was turning black from the wet concrete underneath. More often than not, in an estate cottage like this, a tenant has little voice or comeback in any work that is done – particularly in its quality. To the contractor, the tenant is not the customer, and the work is rarely, if ever, checked, even when concerns and frustrations are raised. A wall heater was ordered for our small shower room, but was comically big and meant the toilet couldn't be used. Months

Kite Flying

later, a heated towel rail was bought; but in installing it, the electrician drilled three times through the waste pipe from the upstairs toilet. We didn't discover the source of the worsening damp and smell until weeks later. After another eight months, no one could be found to tile the single pace of floor, so my mum (accomplished and fit, but still, aged 73) did it. Elsewhere, rotten floorboards are replaced by bits of pallet, doors don't fit properly, handles don't match, concrete doesn't set and falls out from newly fitted windows, and old boilers – thrown out decades ago, to be replaced, are put back in – and although safe, are hardly efficient. The upstairs bathroom window rattles in the wind and howls like a trapped, Scooby-Doo ghost.

We had done the girls' shared bedroom up and managed to put in two small desks so that they had somewhere to do their homework and revise. They were difficult ages to be sharing: Evie nearly 16 and Rosie 12. They shared a chest of drawers, but there was no room for a wardrobe. In the end, we made one out of the narrow airing cupboard, among accumulated decades of redundant and mysterious piping and defunct wiring that had never been removed; the walls were partly covered in dusty, cracked plaster and there were no floorboards.

The day had been strangely hazy. Martin was up a ladder, cutting back the jasmine I'd planted on moving here, almost 16 years ago. It had grown through the girls' open windows and entwined itself (with some encouragement) around the curtain rail, bedposts and shelves. Outside, the sun had been glowing an unsettling, halogen-hot red and the air was thick. The combine harvester had roared into 'Home Field' behind the house that morning like a swarm of bees and we'd cheered its progress, even as we'd rushed to pull the sheets in off the washing line and shut the windows. The barley was in and carted away by lunchtime and the windrows of golden straw laid out in heaped, wavy, shining lines. It was not until I'd wiped the inside windowsill down that I recognised more than harvest dust between my fingers and thumb. Accumulated with the blown husks, chalk and barley dust was the fine grittiness of a desert sandstorm from far away – in time it seemed, too. The surfaces outside were covered with a grey-gold dust that I had seen before, long ago, when Martin was in Iraq, when we had one 18-month-old child and I had a first inkling that we might have to leave our tied cottage at Highclere. The dust was thick on the back doormat. I smiled in a reverie as I realised what it was and watched a precious few swallows and house martins snipping through the dense and gritty-gold air, and wondered that the Sahara had come up to meet them, even as they were beginning to think of leaving to cross it. Martin shouted that he had lost his wedding ring; the thick gold band to my white gold one, missing a tiny diamond and worn with care, thin as a sucked Polo mint.

I leaned out like Juliet to search the nests of sparrows. A blackbird ran along the shed roof, a bright ring around its eye – a circle of Romany gold, the rim of a pound coin we never seemed to have. We scrambled down to search.

I thought of the book I'd sent him, in the Iraqi desert, to keep him safe from harm, where the hot sun melted the glue of the binding, and I thought of all the letters that we wrote. The gold dust fell through the air like glitter, silica twinkling like the tiny flint particles that sparkled up from his feet when we went swimming through the chalk stream to celebrate our tenth wedding anniversary, ten years ago. I found the ring then, nestled by the back door, beneath the jasmine and all the sandstorm-harvest dust. A perfect circle that held me and let me peer into the blank-dark well of life without this; us. The dew-pond centre of the blackbird's eye. For a moment, I didn't pick it up. Like I don't pick up the phone sometimes when it rings too loudly in the dark, quiet house and didn't answer those who called for me in the snowy dark, that night I lost myself on the hill. I could almost feel myself pitching forward; an Alice in Wonderland free fall, grasping at roots. Looking up, for a moment, I saw faint stars above the hill. I paused before picking up the ring. I wanted to feel exactly what it weighed when I gave it back to him on my outstretched palm, like sugar for a horse.

I wanted to savour the moment of feeling safe again here, with us all together in this place. I felt keenly, then, a hopeless gratitude for the continued existence of the Estate and its farm workers' cottages; their peeling estate paint and loose gutters – because that was whom they were built for and for whom they remain. Rural housing: small, functional, hopeful. Not much more than a two-up two-down, square-roomed, council-house style, with a central (often sulky) wood burner, a leaky chimney and a utility room built out of the old woodshed lean-to. Not old enough to be chocolate-box twee and therefore pricey, but built for the dairyman and his family, nine years after the Second World War had ended, within walking distance of the milking parlour, mixed farmyard and a small herd of Jersey milkers. Where else would we live? The Estate continues to provide housing for rural workers, where there is no other, and they are good landlords. Our son must fold his great, long, grasshopper limbs up and into his tiny box bedroom and our girls must continue to share their ever-shrinking space. Meanwhile, we have what is beyond the doorstep to stretch, shake and spread our wings. And there was good news on that front – the former estate manager, who had instigated conservation management on the place, was back.

Outside, we entered a golden place. Spilt grain lay like nubs of treasure on the field corners, for a handful of yellowhammers and sparrows to glean. Freshly cropped stubble fields prompt an unofficial and gleeful socialism of the landscape. We are permitted to walk, run and gallop on it. We let a barely respectable amount of time pass after the grain is carted – and begin our 'work', furtively filling old grain sacks with straw for the chickens and rabbit and running in with a wheelbarrow, for the horses' field shelter. We hailed and whooped our neighbours, doing the same on the other side of the field – the other paramedic in the village, a housekeeper, a cook, a carer – a sort of modern gleaning for us Estate cottagers.

The long contour lines of piled straw, ejected by the combine, echo the shape of the hill; the curves and straight lines ancient and modern, lined up like steeplechase fences; straw, equidistant hedges. In a harvest tradition I've kept almost all my life and the children, theirs, we lined up in a race to jump the windrows. The dog flew ahead of us, remembering this annual game, and we ran and leapt until we fell into the straw piles, breathless and laughing. We returned to fluff and pat them back into shape. And then we took our aged and battered kite to the broad headland below the big down and launched it among the buzzards and the real red kites, finely balanced on the pivots and subtle wobbles of their delta tails. I flew the kite first, but was brought up suddenly by exquisite and unexpected birdsong the way a much-loved tune suddenly connects above the noisy hubbub of a café, when you hadn't been aware of any music playing. The kite faltered and Martin grabbed the string. I stepped aside and cupped my

Kite Flying

ears to the sky, straining to see and hear. There. Undeniably, gloriously, a woodlark singing. And I felt like my heart might break with the joy of it. *'Lullula arborea'*, a song more plaintive than the brassy brightness of a skylark. The descending cascade of notes can sound like stolen snatches of nightingale song, a nuthatch piping or a mistle thrush when heard in part. The *'lu, alu, alu, aluias'* on a descending scale are such a swooning lament. A Pre-Raphaelite bird, if ever I heard one. In recent years, a conservation trend of overwintered stubble seemed to have attracted and sustained them; though they are often transient, camping out for a fortnight before moving on. In the lemony, grey-wagtail, harvest-supper light I spotted it, high above the kite like hope, a chunkier skylark, a little ball of fast-fluttering effort, cascading song; a golden snitch or Betjeman's 'Heart of Thomas Hardy', which flew out of Stinsford Churchyard like a little thumping fig and rocketed over the elms. I shouted and leapt and laughed, and there, in the temporary, unwritten freedom of a stubble field, I knew what it all meant and how much. There would always be new kinds of enclosure, destruction and fencing out. And we knew what to do about that. But for now, we celebrated the annual freedom of harvest, my middle daughter with the kite now and her harvest-gold, sand-coloured skin between brother and sister, pale as dog roses on May nights, all bright, jubilant, astonishing, dancing around us like kites on a string.

The End.

Acknowledgements

I've been writing about nature since, it seems, forever. And this book has been at least eight years in the writing. It was written in the gaps between school runs, several part-time jobs, other writing, raising a family, running to the hills and too much time on Twitter. I have a lot of heartfelt thanks to give. To my dearest, wisest and funniest friend Sarah Evans (the Thoroughbred to my New Forest pony) who has cantered beside me all the way: may there always be dogs, birds and somehow, horses – and an occasional otter would be nice – and to Colin, Amy and Emma, thank you for the loan.

To the community of Inkpen, the little village below the big hill, and its joyous support network: the school, the Foxies, the Yellow Pages Book Club and all those in between; here's to the fresh air, not-so-fresh-air, water run-outs, power outages, moonlit walks, sunsets, village hall parties, playing field gatherings, sheep and mud.

Thank you to the community of Hungerford, the John O'Gaunt School 'JOG family', the Library Gang and Emma Milne-White at the fantastic Hungerford Bookshop – the creative and cultural hub of our little town. And to my staffroom writing pal, Hoffi Robinson: what a treat and privilege it is to share, voice and bounce our various writing ideas, hesitations and obstacles around. You are an inspiration and a very talented writer.

Local journalism is at the root of democracy, records the history of place and gives voice to those that may otherwise

Acknowledgements

struggle to be heard. The *Newbury Weekly News* continues to be one of the very best examples, and I am grateful (and proud) to have honed my nature writing there. 'Keep it tight, keep to the deadline'. Thank you for the column inches, as well as the news.

The RSPB was begun by an all-female pressure group in 1891; forgotten about until Tessa Boase's book, *Mrs Pankhurst's Purple Feather* (*Etta Lemon: The Woman Who Saved the Birds*) set the record straight. I am privileged to have been an early female pioneer of the 'new' nature writing, when the great Rob Hume took me on as a columnist for the RSPB's magazine (formerly *Birds*, now *Nature's Home*) back in 2004. Through it, I charted my children's growing up with (and through the lens of) nature and hopefully showed how we can all make a difference – making sure there's always somewhere a bird can return to. Thank you to all at the RSPB. And especial thanks to Andy Hay, photographer, who joined us for four days each year to illustrate the columns and the 'how to's'. These daft capers are among our best memories of the children's childhood. Your patience knew no bounds.

Without the RSPB, I wouldn't have met the wonderful, warm, funny and generous-spirited author and editor, Derek Niemann. Derek wrote the astonishing *Birds in a Cage*, detailing how four prisoners of war endured captivity and went on to lead and modernise conservation after the war, including at the RSPB. I initially wrote for Derek when he was editor for the RSPB's Youth Magazines. He put me through my paces, stretched and challenged me and put other opportunities in my path. He is an absolute mentor and none of this would have happened without his encouragement and belief in me. Thank you, Derek.

A huge and heartfelt thank you to all the staff, past and present, at Newbury Library – and libraries and librarians everywhere; you are more important now than ever before. And to Helen Dahlke and Diane Coulson for all the writerly lunches that helped get this book off the ground in the first place.

Thank you to my lovely agent Anne Williams at the Kate Hordern Literary Agency (who my youngest still pictures wearing a superhero cape), and for the London train that drew up unwontedly that warm summer's day at our tiny rural station: you were the only person to get out; but you had seen something in my words and we celebrated, up with the linnets on the downs, and you've been patiently waiting, nurturing and advising me ever since. Thank you, truly, for your faith, belief and wisdom. This book happened because of you.

Wholehearted thanks for the heart and home of my publishers, Chelsea Green. To Margo Baldwin, Matt Haslum and Muna Reyal who also saw something in a rangy book that could not easily be put into a neat box. What, in life, can be? Thank you to Rosie Baldwin, Katie Read and the rest of the CGP team, too. Thank you all for really 'seeing me', for making that leap of faith and doing so in such a nurturing, warm and mutually conversational way. I love the ethos, values and environmental standards Chelsea Green stand by and know this book has found its right home. Thank you to my editor, Muna Reyal, whose involved sensitivity, insight, vision, knowledge and warmth has made the editorial process such an exciting pleasure. Any errors are, of course, my own.

I owe so much to the wild and bookish Twitter community: for me, it has been a source of delight, learning, sharing, inspiration, campaigning and uplifting positivity. Thank you all for your humour, kindness, warmth and good-heartedness. I count you as very real friends.

Acknowledgements

To those that protest a wrong: done peacefully, with love and humanity, it is always worth it. It's how we change for the better and learn to listen. Someone or something will benefit down the line. Never feel you haven't made a difference.

Particular thanks to the following people for the encouragement and support they have shown me: to Melissa Harrison for her kind good sense, wisdom, generosity of spirit and all-round creative inspiration; to Amy-Jane Beer for being such a great wild woman advocate and always giving a hand up; to Brigit Strawbridge Howard for her love and support; and to other kind and gentle (but fiercely loving) women too – Jo Cartmell, Katharine Norbury, Sally Goldsmith, Annie O'Garra Worsley, Rachel Malik and Tanya Shadrick. To Jonathan Pomroy (it's your hill too!), Kit Jewitt, Dominic Couzens, James Lowen, Dara McAnulty, Fergus Collins, Ben Hoare, Stephen Moss, and Ollie Douglas at The MERL, and to Tim Dee and Simon Spanton, who said such wonderful words about my writing that I shall take to the end of my days.

Thank you to dear Nic Wilson, writer, fellow *Guardian* Country Diarist and kindred spirit, for her support and all-important encouragement during a 'writerly wobble' at New Networks for Nature, 2019.

And to Ginny Battson, symbioethicist, fluminist and ecophilosopher, for giving us the ecoliteracy we need now and are going to need in the decades ahead, and for the idea of love and ecology.

And to my family. To Sue and Roger Chester for their absolute love and encouragement, to the Allens, the Sayers, my dear brother Ian and the Aussie Sumpters, the Lees and our Summerly.

To my lovely, fun, independent and spirited mum, for all the wildness you instilled in me and all the book covers you wrote in for me, thank you. To Billy, Evie and Rosie, I am so very proud of the creative and principled young people you are; I can't wait to see what happens next!

To Martin, my soulmate and adventurer, who has been there all along, even before you actually were. Thank you for your patience, calm sense and humour, for being such a great dad to our children, for picking up where I leave off, for understanding when I need to 'go to earth' and for being the hill I call home.

And to my dear dad. You'd have been amazed – and slightly bemused at what I've done – and wondered where I'd 'got it from'. Much of it, was you. *x*

About the Author

Nicola's clarion call to writing about nature as a resistance to its loss was first recognised when she won the *BBC Wildlife Magazine*'s 'Nature Writer of the Year Award' in 2003. She has written a regular column for the RSPB members' magazine, *Nature's Home*, ever since and is a *Guardian* Country Diarist. She wrote the first book in the RSPB's Spotlight series on iconic British wildlife, *Otters,* and her writing features in several anthologies, including the *Seasons* books, edited by Melissa Harrison, *Red Sixty Seven,* edited by Kit Jewitt, and *Women on Nature,* edited by Katharine Norbury.

Nicola is a secondary school librarian and runs the school Environment Club. She has a growing following on Twitter (@nicolawriting) and through her blog nicolachester.wordpress.com, writing about the wildlife that surrounds her tenanted rural cottage, where she lives with her family below 'the big hill'. From her writing hut, she can sometimes hear a woodlark singing. And a barn owl has been known to land on her roof.